# Membrane Systems
# for Wastewater Treatment

## Water Environment Federation

**WEF Press**
McGraw-Hill

New York  Chicago  San Francisco  Lisbon  London  Madrid
Mexico City  Milan  New Delhi  San Juan  Seoul
Singapore  Sydney  Toronto

## The McGraw·Hill Companies

Cataloging-in-Publication Data is on file with the Library of Congress.

Copyright © 2006 by the Water Environment Federation. All rights reserved. Printed in the United States of America. Except as permitted under the United States Copyright Act of 1976, no part of this publication may be reproduced or distributed in any form or by any means, or stored in a data base or retrieval system, without the prior written permission of the Water Environment Federation.

*Water Environment Research, WER,* and *WEFTEC* are registered trademarks of the Water Environment Federation.

1 2 3 4 5 6 7 8 9 0 DOC/DOC 0 1 0 9 8 7 6 5

ISBN 0-07-146419-0

*The sponsoring editor for this book was Larry S. Hager and the production supervisor was Pamela A. Pelton. It was set in Sabon by Lone Wolf Enterprises, Ltd. The art director for the cover was Anthony Landi.*

Printed and bound by RR Donnelley.

 This book was printed on recycled, acid-free paper containing a minimum of 50% recycled, de-inked fiber.

McGraw-Hill books are available at special quantity discounts to use as premiums and sales promotions, or for use in corporate training programs. For more information, please write to the Director of Special Sales, McGraw-Hill Professional, Two Penn Plaza, New York, NY 10121-2298. Or contact your local bookstore.

---

**IMPORTANT NOTICE**

The material presented in this publication has been prepared in accordance with generally recognized engineering principles and practices and is for general information only. This information should not be used without first securing competent advice with respect to its suitability for any general or specific application.

The contents of this publication are not intended to be a standard of the Water Environment Federation (WEF) and are not intended for use as a reference in purchase specifications, contracts, regulations, statutes, or any other legal document.

No reference made in this publication to any specific method, product, process, or service constitutes or implies an endorsement, recommendation, or warranty thereof by WEF.

WEF makes no recommendation or warranty of any kind, whether expressed or implied, concerning the accuracy, product, or process discussed in this publication and assumes no liability.

Anyone using this information assumes all liability arising from such use, including but not limited to infringement of any patent or patents.

## *Water Environment Federation*

Founded in 1928, the Water Environment Federation (WEF) is a not-for-profit technical and educational organization with members from varied disciplines who work toward the WEF vision of preservation and enhancement of the global water environment. The WEF network includes water quality professionals from 76 Member Associations in 30 countries. For information on membership, publications, and conferences, contact

Water Environment Federation
601 Wythe Street
Alexandria, VA 22314-1994 USA
(703) 684-2400
www.wef.org

# Contents

List of Tables . . . . . . . . . . . . . . . . . . . . . . . . . . . . . . . . . . . . . . . . . . . . . . . . xv
List of Figures . . . . . . . . . . . . . . . . . . . . . . . . . . . . . . . . . . . . . . . . . . . . . . xvii
Preface . . . . . . . . . . . . . . . . . . . . . . . . . . . . . . . . . . . . . . . . . . . . . . . . . . . . xxi

## Chapter One: Introduction

History of Membrane Treatment for Wastewater . . . . . . . . . . . . . . . . . . . . . 2
Current Trends in Membrane Applications for Wastewater Treatment . . . . . 3
    Wastewater Reclamation . . . . . . . . . . . . . . . . . . . . . . . . . . . . . . . . . . . . . 5
    Reuse Regulations and Technologies . . . . . . . . . . . . . . . . . . . . . . . . . . . . 7
        Industrial Makeup Water . . . . . . . . . . . . . . . . . . . . . . . . . . . . . . . . . 9
        Groundwater Recharge . . . . . . . . . . . . . . . . . . . . . . . . . . . . . . . . . 10
    Membrane Applications . . . . . . . . . . . . . . . . . . . . . . . . . . . . . . . . . . . . 10

## Chapter Two: Membrane Equipment and System Overview

Introduction . . . . . . . . . . . . . . . . . . . . . . . . . . . . . . . . . . . . . . . . . . . . . . 16
Membrane Operational Terminology . . . . . . . . . . . . . . . . . . . . . . . . . . . . 16
Membrane Classifications . . . . . . . . . . . . . . . . . . . . . . . . . . . . . . . . . . . . 18
Terminology for Membrane System Components . . . . . . . . . . . . . . . . . . . 19

Membrane Materials and Configurations . . . . . . . . . . . . . . . . . . . . . . . . . . . 23
Membrane System Configurations and Components . . . . . . . . . . . . . . . . . . 34
    Pretreatment Requirements . . . . . . . . . . . . . . . . . . . . . . . . . . . . . . . . 37
    Flow Equalization . . . . . . . . . . . . . . . . . . . . . . . . . . . . . . . . . . . . . . . 37
    Fine Screening . . . . . . . . . . . . . . . . . . . . . . . . . . . . . . . . . . . . . . . . . 37
    Grit Removal . . . . . . . . . . . . . . . . . . . . . . . . . . . . . . . . . . . . . . . . . . 39
    Fats, Oils, and Grease Removal . . . . . . . . . . . . . . . . . . . . . . . . . . . . 39
    Membrane System Equipment Components . . . . . . . . . . . . . . . . . . . . 40
        Membranes and Associated Hardware . . . . . . . . . . . . . . . . . . . . 40
        Permeation System . . . . . . . . . . . . . . . . . . . . . . . . . . . . . . . . . . 41
General Membrane Operations and Maintenance Procedures . . . . . . . . . . . 41
    Clean-In-Place System . . . . . . . . . . . . . . . . . . . . . . . . . . . . . . . . . . . 43
    Monitoring Systems . . . . . . . . . . . . . . . . . . . . . . . . . . . . . . . . . . . . . 44
    Typical Instrumentation and Controls . . . . . . . . . . . . . . . . . . . . . . . 48
    Transmembrane Pressure Alarms . . . . . . . . . . . . . . . . . . . . . . . . . . . 48
    Backwash . . . . . . . . . . . . . . . . . . . . . . . . . . . . . . . . . . . . . . . . . . . . . 49
    Posttreatment Systems . . . . . . . . . . . . . . . . . . . . . . . . . . . . . . . . . . . 53

# Chapter Three: Membrane Bioreactors

Introduction . . . . . . . . . . . . . . . . . . . . . . . . . . . . . . . . . . . . . . . . . . . . . . . . 58
    Applications . . . . . . . . . . . . . . . . . . . . . . . . . . . . . . . . . . . . . . . . . . . 59
    Water Quality . . . . . . . . . . . . . . . . . . . . . . . . . . . . . . . . . . . . . . . . . 62
        Influent Quality . . . . . . . . . . . . . . . . . . . . . . . . . . . . . . . . . . . . . 62
        Effluent Quality . . . . . . . . . . . . . . . . . . . . . . . . . . . . . . . . . . . . 63
    Treatment Mechanism . . . . . . . . . . . . . . . . . . . . . . . . . . . . . . . . . . . 64
    Equipment Manufacturers . . . . . . . . . . . . . . . . . . . . . . . . . . . . . . . . 65
Process Configurations . . . . . . . . . . . . . . . . . . . . . . . . . . . . . . . . . . . . . . . 66
    Pretreatment Requirements . . . . . . . . . . . . . . . . . . . . . . . . . . . . . . . 66
        Fine Screening . . . . . . . . . . . . . . . . . . . . . . . . . . . . . . . . . . . . . 66
        Primary Clarification . . . . . . . . . . . . . . . . . . . . . . . . . . . . . . . . 67

|    Biological Treatment ............................................ 67
        Recycle of Mixed Liquor ................................... 70
        Solids Retention Time ..................................... 71
        Quality of Activated Sludge ................................ 72
        Mixed Liquor Suspended Solids Concentration ................ 72
        Oxygen Transfer ........................................... 73
    Membrane Separation .......................................... 73
    Cleaning and Chemical Feed Systems ............................ 74
    Posttreatment ................................................ 75
    Residuals Disposal ........................................... 75
Equipment Configurations ........................................... 76
    Membrane Bioreactor Facility Components ....................... 76
    Bioreactor Equipment ......................................... 76
        Biological Process Blowers ................................ 76
        Biological Process Aeration ............................... 77
        Mixed Liquor Recirculation Pumps .......................... 78
        Mixers .................................................... 80
    Membrane Filtration Equipment ................................ 80
        Membrane Configurations ................................... 80
        Permeation System ......................................... 81
        Backpulse Pumps ........................................... 83
        Backpulse Tanks ........................................... 83
        Cleaning System ........................................... 84
        Membrane Air Scour Blowers ................................ 85
        Drain Pumps ............................................... 86
        Air Compressors for Pneumatic System Components ........... 87
        Staging Tank .............................................. 87
    Instrumentation and Controls ................................. 87
        Programmable Logic Controllers ............................ 87
        Turbidimeters ............................................. 88
        Dissolved Oxygen Meters ................................... 88
Operation ......................................................... 89
    Pretreatment ................................................. 89

- Biological Process Operation .................................. 89
  - Sludge Wasting ........................................... 90
  - Dissolved Oxygen ......................................... 91
  - Alkalinity and pH ........................................ 91
- Membrane Filtration—Modes of Operation ....................... 92
  - Permeation ............................................... 92
  - Relax .................................................... 92
  - Backpulse/Backwash ....................................... 93
  - Chemically Enhanced Backpulse/Backwash (Maintenance Clean) .... 93
  - Recovery Cleaning ........................................ 94
  - Flow Control ............................................. 94
- Membrane Aeration ............................................ 95
- Alarms and Trends ............................................ 95
- Operational Changes as a Result of Events and Seasonal Changes ...... 96
- Capacity Management .......................................... 96
- Optimization ................................................. 96

Routine Monitoring ............................................... 97
- Pretreatment ................................................. 97
- Biological Treatment ......................................... 97
- Membrane Systems ............................................. 99
  - System Performance ....................................... 99
  - Equipment and Instrumentation ............................ 100

Maintenance ...................................................... 101
- Cleaning ..................................................... 101
  - Maintenance Cleaning ..................................... 101
  - Recovery Cleaning ........................................ 103
  - Physical Cleaning ........................................ 106
- Membrane Integrity ........................................... 106
  - Integrity Testing ........................................ 106
  - Identifying and Repairing Damaged Membrane Elements ...... 107
- Instrument Calibration ....................................... 108

Troubleshooting .................................................. 109

# Chapter Four: Low-Pressure Membranes for Effluent Filtration

Introduction ................................................... 115
   Applications ............................................... 115
   Feed Water Quality ......................................... 116
   Filtrate Water Quality ..................................... 117
   Treatment Mechanism ........................................ 118
   Safety ..................................................... 120
Process Configurations ........................................ 120
   Pretreatment ............................................... 121
   Membrane System ............................................ 123
   Chemical Storage and Feed Systems .......................... 124
   Posttreatment .............................................. 124
   Residuals Handling ......................................... 125
Equipment Configurations ...................................... 126
   Pretreatment Components .................................... 126
   Membrane Equipment Components .............................. 128
Operation ..................................................... 128
   Pretreatment ............................................... 129
   Automated Systems .......................................... 131
      Filtrate Flow Control .................................. 132
      Transmembrane Pressure Alarms .......................... 133
      Backwash System ........................................ 133
   Productivity Management .................................... 133
      Pretreatment ........................................... 135
      Design Operating Flux .................................. 135
      Backwashing Frequency .................................. 136
      Clean-In-Place Sequence ................................ 136
      Seasonal Changes in Operation .......................... 137
Routine Monitoring ............................................ 137

| | |
|---|---|
| Pretreatment System | 137 |
| Membrane System | 138 |
| Test Parameters | 138 |
| Data Analysis and Reporting | 139 |
| Maintenance | 140 |
| Chemical Cleanings | 141 |
| Integrity Testing | 141 |
| Tank Washing and Cleaning | 142 |
| Instrument Calibration | 144 |
| Troubleshooting | 145 |

# Chapter Five: Nanofiltration and Reverse Osmosis for Advanced Posttreatment

| | |
|---|---|
| Introduction | 149 |
| Applications | 149 |
| Water Quality | 150 |
| Treatment Mechanism | 151 |
| Equipment Manufacturers | 152 |
| Safety | 152 |
| Process Configurations | 152 |
| Pretreatment | 153 |
| Particulate Control | 154 |
| Chlorination/Dechlorination | 154 |
| Scale Prevention | 156 |
| Polyphosphonates | 157 |
| Polyacrylates | 158 |
| Membrane Treatment | 158 |
| Chemical Cleaning Systems | 159 |
| Posttreatment Facilities | 159 |
| Concentrate Disposal | 160 |

| Equipment Configurations | 161 |
|---|---|
|    Membrane Process Equipment Components | 162 |
|       Two- and Three-Stage Membrane Array (Concentrate Staged) | 162 |
|       Two-Pass or Partial Two-Pass Membrane Array (Permeate Staged) | 163 |
|       Membrane Elements | 163 |
|    Typical Instrumentation and Controls | 164 |
| Operation | 164 |
|    Pretreatment Chemistry | 165 |
|    Fouling | 165 |
|       Sulfate Salts | 166 |
|       Silica | 166 |
|       Organics | 167 |
|    Instrumentation and Controls | 167 |
|       Recovery | 167 |
|       Feed and Differential Pressure | 168 |
|       High and Low pH | 168 |
|    Event-Driven Systems | 168 |
|       Cleaning Events | 169 |
|       Source Water Changes | 170 |
|       Biofilm Development | 170 |
|    Capacity Management | 170 |
|    Process Optimization | 171 |
| Routine Monitoring | 172 |
|    Feed Water | 172 |
|    Membrane System | 174 |
|    Data Interpretation and Reporting | 175 |
|       Percent Rejection | 176 |
|       Mass-Transfer Coefficient (Specific Flux) | 176 |
|       Differential Pressure | 176 |
| Maintenance | 177 |
|    Chemical Cleanings | 177 |
|    Integrity Testing | 178 |
|    Instrument Calibration | 180 |
| Troubleshooting | 181 |

# Chapter Six: Membrane System Case Studies

Introduction .................................................. 185
Case Study No. 1—Running Springs Water Recycling Plant .......... 186
   Background ............................................... 186
   Process Overview .......................................... 188
   Plant Data and Design Summary ............................. 190
   Highlights ................................................ 191
Case Study No. 2—The Hamptons Water Recycling Facility .......... 193
   Background ............................................... 193
   Process Overview .......................................... 193
   Plant Data and Design Summary ............................. 196
   Highlights ................................................ 197
Case Study No. 3—Traverse City Regional Wastewater
               Treatment Plant ..................................... 198
   Background ............................................... 198
   Process Overview .......................................... 200
   Plant Data and Design Summary ............................. 202
   Highlights ................................................ 202
Case Study No. 4—Key Colony Water Reuse Plant ................... 202
   Background ............................................... 203
   Process Overview .......................................... 205
   Plant Data and Design Summary ............................. 207
   Highlights ................................................ 207
Case Study No. 5—West Basin Water Recycling Plant ............... 208
   Background ............................................... 208
   Process Overview .......................................... 209
      Microfiltration/Double-Pass Reverse-Osmosis Process
      Description ............................................ 211
      Product Water Quality .................................. 212
   Plant Data and Design Summary ............................. 212
   Highlights ................................................ 212

Case Study No. 6—Chandler, Arizona, Reverse-Osmosis Facility
        Wastewater Reclamation Plant .................. 214
   Background ............................................. 214
   Process Overview ....................................... 216
   Plant Data and Design Summary ......................... 218
   Highlights ............................................. 220
Case Study No. 7—Scottsdale Water Campus ..................... 222
   Background ............................................. 222
   Process Overview ....................................... 222
      Microfiltration .................................... 224
      Reverse Osmosis ................................... 224
      Posttreatment Facilities ........................... 225
   Plant Data ............................................. 226
   Highlights ............................................. 227

# Glossary ................................................................. 229

# Appendix I: Related Equations ................. 241

# Appendix II: References and Resources ....... 251

# Index ................................................... 261

# List of Tables

| | | |
|---|---|---|
| 1.1 | U.S. Membranes Development Timeline | 4 |
| 1.2 | Summary of U.S. EPA Guidelines for Water Reuse | 8 |
| 1.3 | Common Membrane Applications by Type of Reuse | 9 |
| 1.4 | Applications for Membrane Technologies in Wastewater Treatment | 11 |
| 2.1 | General Characteristics of Membranes | 19 |
| 2.2 | Membrane Material Characteristics | 26 |
| 2.3 | Advantages and Disadvantages of Various Membrane Materials | 27 |
| 2.4 | Summary of Key Features of Selected MF and UF Membranes | 35 |
| 2.5 | Summary of Integrity Monitoring Methods | 46 |
| 3.1 | Typical Municipal MBR Effluent Quality | 64 |
| 3.2 | Typical MBR System Testing Requirements | 98 |
| 3.3 | Recovery Cleaning Chemical Selection Chart | 105 |
| 3.4 | Membrane Bioreactor Troubleshooting | 110 |
| 4.1 | Typical Feed Water (Secondary Effluent) Quality | 116 |
| 4.2 | Typical Filtrate Water Quality for MF and UF Treatment Facilities | 118 |
| 4.3 | Apparent Dimensions of Small Particles, Molecules, and Ions | 119 |
| 4.4 | Typical Characteristics of Chemical Cleaning Solutions in Low-Pressure Membranes | 126 |
| 4.5 | Typical Opening/Pore Sizes for MF/UF Membranes and Common Pretreatment Devices | 127 |
| 4.6 | Minimum Testing Requirements for MF and UF Facilities | 139 |

| | | |
|---|---|---|
| 4.7 | Typical Membrane Integrity Test Times | 142 |
| 4.8 | General Troubleshooting Guidelines for MF and UF Facilities | 145 |
| 5.1 | Typical Feed Water Quality Ranges for RO Treatment Facilities | 150 |
| 5.2 | Typical Product Water Quality Ranges for RO Treatment Facilities | 151 |
| 5.3 | Typical List of Alarms for a RO System | 169 |
| 5.4 | Reverse-Osmosis/NF System Monitoring Requirements | 175 |
| 5.5 | Reverse-Osmosis/NF Membrane Process Troubleshooting Guide | 181 |
| 6.1 | Running Springs Water Recycling Plant | 186 |
| 6.2 | Running Springs Hydraulic Performance Data | 190 |
| 6.3 | Running Springs Membrane Filtration Process Design Parameters | 192 |
| 6.4 | Running Springs Treatment Results/Requirements | 192 |
| 6.5 | The Hamptons Water Recycling Facility | 195 |
| 6.6 | The Hamptons Membrane Filtration Process Design Parameters | 197 |
| 6.7 | The Hamptons Treatment Results and Permit Requirements | 198 |
| 6.8 | Traverse City Regional Wastewater Treatment Plant | 199 |
| 6.9 | Traverse City Membrane Filtration Process Design Parameters | 203 |
| 6.10 | Traverse City Treatment Results and Permit Requirements | 203 |
| 6.11 | Key Colony Water Reuse Plant | 204 |
| 6.12 | Key Colony Membrane Filtration Process Design Parameters | 207 |
| 6.13 | Key Colony Treatment Results and Permit Requirements | 208 |
| 6.14 | West Basin Water Recycling Plant | 209 |
| 6.15 | Effluent Water Quality from the Hyperion Treatment Plant | 213 |
| 6.16 | First- and Second-Pass RO Product Water Quality | 213 |
| 6.17 | Chandler Industrial–Municipal Water Reclamation Plant | 215 |
| 6.18 | Chandler RO Facility Effluent Water Quality | 218 |
| 6.19 | Chandler Field Modifications | 220 |
| 6.20 | Typical Chandler Operating Conditions in 2004 | 221 |
| 6.21 | Scottsdale Water Campus | 223 |
| 6.22 | Advanced Water Treatment Facility Influent Water Quality | 224 |
| 6.23 | Advanced Water Treatment Facility Effluent Water Quality | 226 |

# List of Figures

| | | |
|---|---|---|
| 1.1 | Schematic Diagram of Osmosis and RO | 3 |
| 1.2 | Decentralized Reuse Schematic | 7 |
| 1.3 | Overview of Membrane Flow Schematics | 12 |
| 2.1 | Comparison of Membrane Filtration Processes | 20 |
| 2.2 | One Frame of Pressurized MF Membrane Modules | 22 |
| 2.3 | Several Frames of a Pressurized UF System | 23 |
| 2.4 | Cross-Sections of Various Membrane Types | 24 |
| 2.5 | Cross-Section View of a Flat Plate and Frame Membrane | 29 |
| 2.6 | Cross-Section View of an Outside-In Flow Hollow-Fiber Membrane | 30 |
| 2.7 | Photograph of a Submerged Hollow-Fiber Membrane Cassette | 30 |
| 2.8 | Cutaway of an Enclosed Hollow-Fiber UF Module | 31 |
| 2.9 | Cutaway Schematic of an Individual Enclosed Hollow-Fiber Membrane Module | 32 |
| 2.10 | Tubular Membrane Configuration | 32 |
| 2.11 | Components of Spiral-Wound Membrane | 33 |
| 2.12 | Spiral-Wound Membrane Configuration | 33 |
| 2.13 | Typical Pressure Vessel for Spiral-Wound Membranes | 34 |
| 2.14 | Illustration of an Automatic Screen | 38 |
| 2.15 | Chemical Cleaning System at a Low-Pressure Membrane Facility | 44 |
| 2.16 | Immersed Low-Pressure Membrane during Backwash | 51 |
| 3.1 | Conventional Activated Sludge Schematic | 58 |

| | | |
|---|---|---|
| 3.2 | External MBR Schematic | 59 |
| 3.3 | Immersed MBR Schematic | 59 |
| 3.4 | Mixed Liquor and Permeated Water | 63 |
| 3.5 | Membrane Bioreactor—Nitrification Only | 68 |
| 3.6 | Membrane Bioreactor—Total Nitrogen Removal | 69 |
| 3.7 | Membrane Bioreactor—Nitrogen Removal with Chemical Phosphorus Removal | 70 |
| 3.8 | Membrane Bioreactor—Biological Nutrient Removal | 71 |
| 3.9 | Process Flow Diagram of a Typical MBR Facility | 77 |
| 3.10 | Fine-Bubble Aeration System | 78 |
| 3.11 | Recirculation Pumps | 79 |
| 3.12 | Permeate Pumps with Air Separators | 82 |
| 3.13 | Backpulse Pumps | 83 |
| 3.14 | Air Scour System | 85 |
| 3.15 | Turbidimeter | 89 |
| 3.16 | Modes of MBR Operation | 93 |
| 4.1 | Use of Product Water from Existing Full-Scale Wastewater Treatment Facilities using Membranes | 115 |
| 4.2 | Process Flow Schematic for Effluent Membrane Filtration | 121 |
| 4.3 | Pretreatment Strainers in an Effluent Membrane Filtration System | 122 |
| 4.4 | Schematic Diagram of a Typical Effluent Membrane Filtration System | 123 |
| 4.5 | Illustration of an Immersed Low-Pressure Membrane System | 129 |
| 4.6 | Pressurized MF Treatment System | 130 |
| 4.7 | Cutaway Diagram of a Typical Strainer | 132 |
| 4.8 | Typical Operating Pressures for Low-Pressure Membranes Filtering Secondary Effluent | 134 |
| 4.9 | Pressure Decay Rates in Pressurized UF Frames with One Broken Fiber and with No Breaks | 143 |
| 4.10 | Photograph of Immersed Membrane Tank Drains | 144 |
| 5.1 | Reverse-Osmosis Membrane Cleaning Schematic | 159 |
| 5.2 | Concentrate Staged Array | 162 |
| 5.3 | Permeate Staged Array | 163 |

| | | |
|---|---|---|
| 5.4 | Silt Density Index Apparatus | 174 |
| 5.5 | Permeate Probing Apparatus for a Spiral Wound Membrane Element | 179 |
| 6.1 | Case Study Navigational Chart | 185 |
| 6.2 | Running Springs Retrofit to a MBR | 187 |
| 6.3 | Process Flow Diagram for the Running Springs WRF | 188 |
| 6.4 | Flux and Permeability with Flux Enhancer at 13.3°C | 191 |
| 6.5 | The Hamptons Water Reclamation Facility | 194 |
| 6.6 | Nitrogen Data from the Hamptons WRF | 195 |
| 6.7 | Mixed Liquor Suspended Solids Data from the Hamptons WRF | 196 |
| 6.8 | The Traverse City MBR Plant | 201 |
| 6.9 | Process Flow Diagram for the Key Colony Water Reuse Plant | 205 |
| 6.10 | Key Colony Water Reuse Plant | 206 |
| 6.11 | Overview of Chandler Wastewater Reclamation Plant | 215 |
| 6.12 | Process Flow Diagram for the Chandler Wastewater Reclamation Plant | 217 |
| 6.13 | Process Flow Diagram for the Scottsdale Water Campus | 223 |
| AI.1 | Example Normalized Mass-Transfer Coefficient Plot for a Pressure MF System | 243 |
| AI.2 | Schematic Depicting Percent Recovery | 245 |

# Preface

Welcome to the Water Environment Federation's (WEF's) publication, *Membrane Systems for Wastewater Treatment!* This publication is a comprehensive educational text for professionals concerned with membrane equipment, systems, and processes in wastewater treatment and reclamation.

*Membrane Systems for Wastewater Treatment* provides a comprehensive overview of the various technologies and their applications in the field of wastewater treatment. It is not intended to be a design manual but rather an overview of the system configurations and components as well as general operations and maintenance procedures. The Federation recognizes the huge potential of membrane applications for wastewater treatment. This publication is intended to put practical information about these systems into the hands of wastewater professionals to promote the assessment of membrane technologies as a potential wastewater treatment alternative.

*Membrane Systems for Wastewater Treatment* discusses some background information on general membrane topics before covering details on wastewater applications. The three primary wastewater membrane applications covered in this publication are

- Membrane bioreactors,
- Low-pressure membranes (microfiltration/ultrafiltration) for effluent filtration, and
- Nanofiltration/reverse osmosis for advanced posttreatment.

The discussion of each of these applications includes common process configurations, unique operational considerations, monitoring and control equipment and strategies, general cleaning and maintenance procedures, and troubleshooting guidance. Case studies are included in the last chapter.

This publication was primarily written to contribute to the resources and information available to wastewater professionals, including operators, maintenance technicians, utility managers, engineers/consultants, and engineering students.

Membrane technologies and applications in wastewater treatment and reclamation are evolving. Rapid progress, including the development of new membrane materials, configurations, and operational strategies, makes this an exciting time to work with these systems. The Federation recognizes that this material will require review and update more frequently than many other wastewater topics. We urge the users of this manual to submit any comments and suggested updates for inclusion in future editions of this publication. These comments will be gratefully accepted at optrain@wef.org.

*Membrane Systems for Wastewater Treatment* was produced under the direction of Khalil Atasi, Chair. The principal authors of the publication are

> Khalil Atasi, Wade-Trim, Inc.
>
> George Crawford, CH2M HILL
>
> Jill Manning Hudkins, Hartman & Associates, Inc., A Tetra Tech Co.
>
> Dennis Livingston, Enviroquip, Inc.
>
> Rod Reardon, Camp Dresser & McKee, Inc.
>
> Hal Schmidt, Brown and Caldwell

The following reviewers and contributors are gratefully acknowledged:

> Kevin Alexander
>
> Elena Bailey
>
> Scott Blair
>
> Bob Bucher
>
> Stephen Chapman
>
> Pierre Côte
>
> Ronald Crites
>
> Glen Daigger

Francis DiGiano
Don Dodson
Scott Freeman
Jim Hagstrom
Colin Hobbs
Simon Judd
Jarrett Kinslow
Angela Klein
John Koch
Terry Krause
Henryk Melcer
Trevor Miller
Mehul Patel
Deo Phagoo
Keith Radick
Kim Riddell
Marie-Laure Rodde-Pellegrin
Cory Schneider
Alan Scrivner
Reza Shamskhorzani
Tom Stephenson
Paul Sutton
George Tchobanoglous
Steve Tedesco
Srinivas Veerapaneni
William Vernon
Cindy Wallis-Lage
Fufang Zha

The following organizations provided materials and/or supported author or reviewer efforts:

AES Engineering
American Water Works Association
Black & Veatch Corporation
Bio-Microbics, Inc.
Brown and Caldwell
Camp Dresser & McKee, Inc.
Carollo Engineers
CH2M HILL, Inc.
City of Delphos, Ohio
City of Scottsdale, Arizona
Consept
Drinking Water Inspectorate
Department of Environmental Sciences and Engineering, University of North Carolina at Chapel Hill
Earth Tech
Eimco (Jones and Atwood, Ltd.)
Enviroquip, Inc.
Floyd Browne Group
General Electric
Hach Company
Hartman & Associates, Inc., a Tetra Tech Company
Hydranautics
ITT Sanitaire Americas
IWA Publishing, Ltd., London, United Kingdom
Jordan Jones & Goulding
King County Department of Natural Resources and Parks

Koch Membrane Systems

Kubota Corporation

Malcolm Pirnie, Inc.

Memcor Products, USFilter

OMI

Orange County Water District

Pall Corporation

Plenum Press

P. M. Sutton and Associates

Running Springs Wastewater Treatment Plant

School of Water Sciences, Cranfield University

Separation Processes, Inc.

Tetra Tech, Inc.

The Dow Chemical Company

The Drinking Water Inspectorate, Department for Environment, Food and Rural Affairs

The McGraw-Hill Companies

USFilter

Wade-Trim, Inc.

Water Environment Research Foundation

ZENON Environmental, Inc.

CHAPTER ONE

# Introduction

History of Membrane Treatment for Wastewater . . . . . . . . . . . . . . . . . . . . . . . . . . . 2
Current Trends in Membrane Applications for Wastewater Treatment . . . . . . . . . . . 3
    Wastewater Reclamation . . . . . . . . . . . . . . . . . . . . . . . . . . . . . . . . . . . . . . . . . . 5
    Reuse Regulations and Technologies . . . . . . . . . . . . . . . . . . . . . . . . . . . . . . . . . 7
        Industrial Makeup Water . . . . . . . . . . . . . . . . . . . . . . . . . . . . . . . . . . . . . . 9
        Groundwater Recharge . . . . . . . . . . . . . . . . . . . . . . . . . . . . . . . . . . . . . . 10
    Membrane Applications . . . . . . . . . . . . . . . . . . . . . . . . . . . . . . . . . . . . . . . . 10

## History of Membrane Treatment for Wastewater

In 1748, a French physicist named Jean Antoine Nollet first documented a phenomenon known today as osmosis—the process by which water diffuses through a semipermeable membrane, moving from a liquid solution with fewer contaminants to a liquid solution with more contaminants until equilibrium is reached. Two hundred years later, researchers manipulated this phenomenon to create the first reverse-osmosis system. They found that by applying energy (in the form of pressure or a vacuum) to the liquid solution with more contaminants, the water could still move through the membrane, leaving the contaminants behind and discharging clean water (Figure 1.1). In other words, water can be directed under high pressure or in a vacuum through thin membranes to remove even tiny particles such as salts, viruses, pesticides, and most organic compounds.

Reverse-osmosis (RO) systems were the first type of membrane systems to be used in advanced wastewater treatment. These early applications were specific to water reclamation/reuse and groundwater recharge and were limited geographically to areas facing water shortages. Reverse-osmosis systems can be used to remove soluble ions, dissolved solids, and organic materials from high-quality tertiary effluent to polish final effluents for reuse or for groundwater recharge.

Several additional types of membranes have been developed. The classifications of membranes primarily used in wastewater treatment include the following:

- Microfiltration (MF),
- Ultrafiltration (UF),
- Nanofiltration (NF), and
- Reverse osmosis (RO).

Shown below in Table 1.1 is a timeline of the history of membrane wastewater treatment in the United States. It is important to note that membranes were used to treat drinking water before wastewater treatment applications became viable.

Tighter environmental regulations coupled with recent developments in membrane-manufacturing technology have made the use of

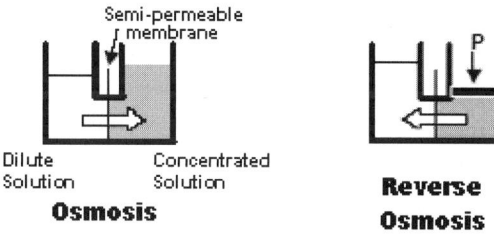

FIGURE 1.1—SCHEMATIC DIAGRAM OF OSMOSIS AND RO.

a broader range of membrane processes more appealing, practical, and economical in the field of wastewater treatment.

# Current Trends in Membrane Applications for Wastewater Treatment

Over the years, wastewater treatment technology advancements have been driven by the need to protect public health and the environment. As populations continue to rise, humans put increasing strain on natural systems. Many communities are scrambling to find solutions to shortages of clean water for direct human consumption, irrigation, and groundwater recharge to mitigate saltwater intrusion issues. The reality is that wastewater treatment effluent is no longer seen as something to just "dispose of"; it is now increasingly viewed as a reliable water source. Time, attention, and research money are being put into developing reliable, cost-effective wastewater treatment processes that can remove a wide range of pollutants and consistently produce high-quality water.

In 1970, the U.S. Environmental Protection Agency (U.S. EPA) was established, followed by the passage of the Clean Water Act (CWA) in 1972. The primary intent of the CWA and subsequent amendments was to improve the waters of the United States so they are safe for contact by humans and able to support diverse healthy aquatic populations. As part of the mandate of the CWA, U.S. EPA is primarily tasked with regulating point-source discharges (municipal and industrial) to U.S. surface waters . In the United States, every discharger must

Table 1.1  U.S. Membranes Development Timeline.*

| 1748 | 1950s | 1970s | 1980s | 1990s | 2000 TO PRESENT |
|---|---|---|---|---|---|
| **NF/RO** | | | | | |
| Osmosis discovered by French Physicist Nollet | First California RO membrane (1960) | Water Factory 21 commissioned in 1976 (California RO, 5 mgd) | Commercialization of NF membranes | West Basin Water Recycling Plant El Segundo, California commissioned in 1997 (MF+RO, 2.5 mgd RO) | |
| | | Commercialization of TFC RO membranes | | Scottsdale, Arizona commissioned in 1999 (MF+RO, 8.5 mgd RO) | |
| **MF/UF AND MEMBRANE BIOREACTORS (MBRs)** | | | | | |
| | Commercialization of MF/UF membranes | Development of external loop MBR technologies | Installed MF/UF capacity on the order of 1.5 mgd as of 1986 | Development of immersed MBR technologies | Installed MF/UF/MBR capacity on the order of 500 mgd as of 2002 |
| | | | Keystone, Colorado Water Treatment Plant; commissioned in 1987 (MF, 0.06 mgd) | Key Colony, Florida and Anthem, Arizona commissioned in 1999 (MBR, 0.9, 0.5 mgd) | Running Springs, California commissioned 2003 (MBR, 0.6 mgd) |
| | | | | | Traverse City, Michigan commissioned 2004 (MBR, 8.5 mgd) |

* mgd × 3785 = m$^3$/d.

have a National Pollutant Discharge Elimination System (NPDES) permit that quantifies what that user can discharge to a given body of water.

Given the mandate of "improving" waters through NPDES permitting, many authorities are forced to reduce the allowable concentration of pollutants in a wastewater stream as water usage in a community increases. For example, if the population of a city doubles, the permitted levels of a contaminant may be cut by one-half to keep the *total maximum daily load* of that contaminant being discharged to a body of water the same. This is a real dilemma for growing communities that are trying to meet increasingly strict effluent requirements. In addition to meeting stricter permit limits for *conventional* and *nonconventional pollutants*, a growing awareness of *emerging pollutants* potentially places new treatment requirements on the regulatory horizon.

- Conventional pollutants—examples are biochemical oxygen demand, total suspended solids, nitrogen, phosphorus, and pathogens;
- Nonconventional pollutants—examples are salts, metals, and refractory organics; and
- Emerging contaminants—examples are pharmaceuticals, household cleaning products, and endocrine-disrupting compounds

Effective wastewater treatment may reduce contaminant loading to streams and other waterways, but it is only one facet of a comprehensive water-management program. Another critical component of any successful program is wastewater reclamation, and membranes are changing this industry at a rapid pace.

# WASTEWATER RECLAMATION

Water reclamation can be described as the treatment, storage, and distribution of wastewater for some kind of beneficial use or reuse application. Reuse is a broad term that can be further characterized as either direct or indirect. As defined by the state of California, *indirect reuse involves the return of highly treated wastewater to a natural environment (groundwater reservoir or stream), where it is mixed or blended with other waters for an extended period of time before being treated via sedimentation, filtration, and disinfection.* (Reclaimed wastewater can also be reused directly for such nonpotable applications as irrigation (e.g., golf courses) or as source water for industrial users (e.g., boiler-water makeup).

In years past, water-reclamation facilities have been typically located within or in close proximity to a wastewater treatment plant sited on the fringes of urban areas. In some cases, potential users of reclaimed water could be found in the vicinity, but expensive transmission systems were typically required to convey reclaimed water to users, making reuse difficult to justify economically.

To address this problem, satellite reclamation systems were installed. This fairly new concept involves the installation of membrane facilities in upstream subwatershed or subcatchment areas to reclaim water for local reuse. By putting smaller membrane facilities in remote areas near the collection system and close to the reuse site, the costs of installing distribution systems to take reclaimed water from the main plant to reuse areas can be avoided (Figure 1.2). Satellite facilities can be of the full-flow type or the sewer-mining type—the latter treating a constant flow from an adjacent sewer and the former treating all flows from the sewer. Several examples of ways in which wastewater can be reused are as follows:

(1) Irrigation of agricultural sites for crop irrigation, tree farms, and commercial nurseries.

(2) Landscape irrigation of golf courses; green spaces such as roadway medians; landscaped areas within commercial, industrial, and/or residential developments; parks; playgrounds; and school grounds.

(3) Industrial uses for cooling water (boiler-water makeup) and process needs.

(4) Groundwater recharge via either spreading basins or direct injection to the groundwater.

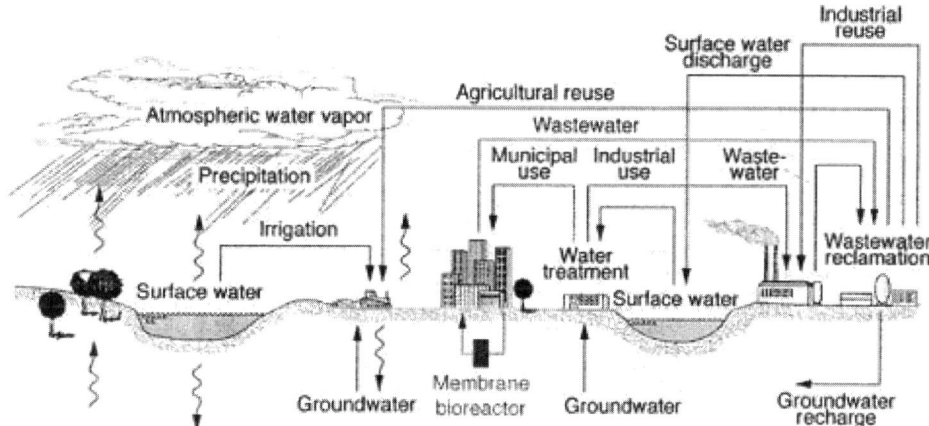

FIGURE 1.2—DECENTRALIZED REUSE SCHEMATIC (COURTESY OF GEORGE TCHOBANOGLOUS).

(5) Recreational and environmental uses such as the development of recreational lakes, marsh enhancement, and streamflow augmentation.

(6) Nonpotable water uses such as fire protection, air conditioning, toilet flushing, construction water, and cleaning sanitary sewers.

(7) Blending water-supply reservoirs and/or augmenting surface water sources for potable water reuse.

The level of treatment required for each type of reuse varies, depending on the assessed risk to public health, and is typically regulated at the state level, with guidance provided by U.S. EPA. The level of treatment required determines what type of membrane system is best suited for the application and, more importantly, the design and operation of the membrane process.

# REUSE REGULATIONS AND TECHNOLOGIES

In 1992, U.S. EPA set guidelines to allow states to develop their own statutory laws regulating water reuse. These guidelines are summarized in Table 1.2 with recommended levels of treatment. As you can see, the level of treatment required is commensurate with the risk to public health. In other words, the more likely someone is to come

Table 1.2  Summary of U.S. EPA Guidelines for Water Reuse.

| TYPE OF USE QUALITY | TREATMENT | RECLAIMED WATER* |
|---|---|---|
| Urban reuse (landscape irrigation) Agricultural reuse Recreational impoundments | Secondary Tertiary filtration Disinfection | pH = 6–9 10 mg/L BOD 2 NTU Non-detect fecal coliform/ 100 mg/L 1 mg/L residual chlorine |
| Restricted access Construction uses Cooling uses | Secondary Disinfection | pH = 6–9 30 mg/L BOD 30 mg/L TSS 200 fecal coliform/100 mL 1 mg/L residual chlorine |
| Groundwater recharge of nonpotable aquifers by spreading/injection | Site specific Primary/secondary | Site specific and use dependent |
| Groundwater recharge of potable aquifers | Site specific Secondary (min.) Disinfection (min.) | Site specific Meet drinking water standards after percolation through vadose zone |
| Augmentation of surface suppliers | Secondary Filtration Disinfection Advanced | pH = 6.5–8.5 2 NTU Non-detect fecal coliform/ 100 mg/L 1 mg/L residual chlorine Meet drinking water standards |

* BOD = biochemical oxygen demand and TSS = total suspended solids.

into contact with the reused water, the higher the level of treatment required. Membranes provide the highest level of treatment recommended by physically rejecting pathogenic microorganisms upstream of any postdisinfection such as UV. In fact, in many states, regulatory requirements are dictated by the ability of membranes to remove *conventional*, *nonconventional*, and *emerging* contaminants from treated wastewater.

The most typical membrane technologies used for indirect and direct reuse are shown in Table 1.3.

Since the early 1980s, aerobic membrane bioreactors (MBR) have been successfully applied to treat a wide range of industrial wastewaters, albeit on a much smaller scale than tubular systems. Applications have included food-processing waste streams, tannery effluents, oily wastes, and landfill leachate. In addition, anaerobic MBRs have been used to treat high-strength wastes, such as dairy wastes and wine-distillery effluents. For illustration purposes, two types of reuse are discussed in more detail below.

Industrial Makeup Water

Industrial uses include water for such purposes as processing, washing, and cooling in manufacturing plants. Industrial uses of reclaimed

Table 1.3  Common Membrane Applications by Type of Reuse.

| Reuse Type Pretreatment* | Description | Membrane Type(s) | Process | Configuration |
|---|---|---|---|---|
| Direct nonpotable | Landscape irrigation | MF/UF | Tertiary-filtration or MBR | Fine screening |
|  | Boiler makeup | RO | AWT | MF/UF/NF |
| Indirect potable | Stream augmentation | MF/UF | Tertiary-filtration or MBR | Fine screening |
|  | Aquifer recharge | MF/UF | Tertiary-filtration or MBR | Fine screening |

* Not all pretreatment requirements listed.

water from external sources include evaporative cooling water, boiler-feed water, process water, and landscaping and maintenance of industrial grounds.

Groundwater Recharge

Groundwater recharge has been used to

(1) Reduce and/or reverse declines in groundwater levels,
(2) Protect underground freshwater in coastal aquifers against saltwater intrusion, and
(3) Store water in the aquifer system for future use.

Direct subsurface recharge is achieved when the effluent is injected directly to the groundwater aquifer. The first water-reuse plant in the United States was Water Factory 21, located in Orange County, California, which was put into service in 1976. This treatment facility was designed to discharge high-quality effluent that would be blended with deep well water and then injected to a coastal hydraulic barrier to prevent saltwater intrusion to the groundwater. This system has proven to be very effective and is still in use today. Water Factory 21 has blended more than 12.5 mil. $m^3$ (33 bil. gal) of treated wastewater with existing groundwater.

Treatment requirements for groundwater recharge vary considerably and depend on a number of items such as

- The purpose of the groundwater recharge,
- Sources of reclaimed water,
- Recharge methods, and
- Location and geological conditions.

# MEMBRANE APPLICATIONS

Table 1.4 contains a fairly comprehensive list of membrane applications for wastewater treatment. All of these applications are not necessarily commonly used today. However, the list demonstrates the range of potential roles that membranes can play in wastewater treatment.

Table 1.4 Applications for Membrane Technologies in Wastewater Treatment (Courtesy of the McGraw-Hill Companies).

| Applications | Description |
|---|---|
| **Microfiltration and Ultrafiltration Systems** | |
| Aerobic biological treatment for discharge or reuse | Membrane is used to separate the treated wastewater from the active biomass in an activated sludge process. The membrane separation unit can be internal immersed in the bioreactor or external to the bioreactor. Such processes are known as MBR processes. |
| Anaerobic biological processes | Membrane is used to separate the treated wastewater from the active biomass in an anaerobic complete-mix reactor. |
| Membrane aeration biological treatment | Plate and frame, tubular, and hollow membranes are used to transfer high purity oxygen to the biomass attached to the outside of the membrane. Such processes are known as membrane aeration bioreactor processes. |
| Membrane extraction biological treatment | Membranes are used to extract degradable organic molecules from inorganic constituents such as acids, bases, and salts from the waste stream for subsequent biological treatment. Such processes are known as extractive membrane bioreactor processes. |
| Effluent reuse | Used to remove residual suspended solids from settled secondary effluent or from the effluent from depth or surface filters to achieve higher effluent quality and/or effective disinfection with either chlorine or UV radiation for reuse applications. |
| Pretreatment for nanofiltration and reverse osmosis | Microfilters are used to remove residual colloidal and suspended solids as a pretreatment step for additional processing. |
| **Nanofiltration** | |
| Effluent reuse | Used to treat prefiltered effluent (typically with microfiltration) for indirect potable reuse applications such as groundwater injection. Credit is also given for disinfection when using nanofiltration. |
| Wastewater softening | Used to reduce the concentration of multivalent ions contributing to hardness for specific reuse applications. |

Table 1.4  Applications for Membrane Technologies in Wastewater Treatment (Courtesy of the McGraw-Hill Companies) *(continued)*.

| APPLICATIONS | DESCRIPTION |
| --- | --- |
| **REVERSE OSMOSIS** | |
| Effluent reuse | Used to treat prefiltered effluent (typically with microfiltration) for indirect potable reuse applications such as groundwater injection. Credit is also given for disinfection when using reverse osmosis. |
| Effluent dispersal | Reverse osmosis processes have proved capable of removing sizable amounts of selected compounds such as N-nitrosodimethylamine. |
| Two-stage treatment for boiler use | Two stages of reverse osmosis are used to produce water suitable for high-pressure boilers. |

This publication will focus on the applications most commonly used today, which are

- Membrane bioreactors with either submerged or external membrane systems (Chapter 3),
- Conventional activated sludge treatment systems with MF/UF effluent filtration (Chapter 4),
- Membrane bioreactors with NF/RO advanced posttreatment (Chapter 5), and
- Conventional activated sludge treatment systems with MF/UF effluent filtration and NF/RO advanced posttreatment (Chapter 5).

Figure 1.3 provides a summary of the various flow schematics that can be used for membrane systems in wastewater treatment applications. Though our knowledge about membrane processes in wastewater treatment applications grows every day, it is still in its infancy. Process designs are often based on pilot testing and experience. Theoretical frameworks to describe various applications and processes in

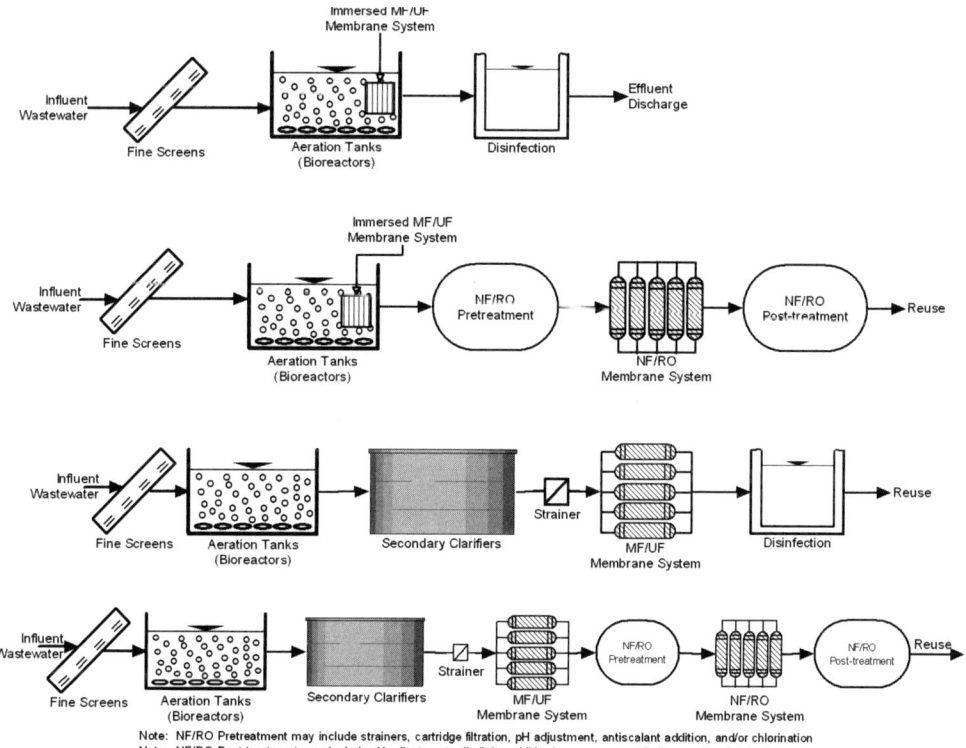

**FIGURE 1.3—OVERVIEW OF MEMBRANE FLOW SCHEMATICS.**

wastewater treatment or to model performance under various conditions are still being researched and studied to develop a more robust scientific and mathematical understanding of these processes.

# CHAPTER TWO

# Membrane Equipment and System Overview

Introduction . . . . . . . . . . . . . . . . . . . . . . . . . . . . . . . . . . . . . . . . . . . . . . . . 16
Membrane Operational Terminology . . . . . . . . . . . . . . . . . . . . . . . . . . . . 16
Membrane Classifications . . . . . . . . . . . . . . . . . . . . . . . . . . . . . . . . . . . . 18
Terminology for Membrane System Components . . . . . . . . . . . . . . . . . . 19
Membrane Materials and Configurations . . . . . . . . . . . . . . . . . . . . . . . . 23
Membrane System Configurations and Components . . . . . . . . . . . . . . . 34
    Pretreatment Requirements . . . . . . . . . . . . . . . . . . . . . . . . . . . . . . . 37
    Flow Equalization . . . . . . . . . . . . . . . . . . . . . . . . . . . . . . . . . . . . . . 37
    Fine Screening . . . . . . . . . . . . . . . . . . . . . . . . . . . . . . . . . . . . . . . . 37
    Grit Removal . . . . . . . . . . . . . . . . . . . . . . . . . . . . . . . . . . . . . . . . . 39
    Fats, Oils, and Grease Removal . . . . . . . . . . . . . . . . . . . . . . . . . . . . 39
    Membrane System Equipment Components . . . . . . . . . . . . . . . . . . 40
        Membranes and Associated Hardware . . . . . . . . . . . . . . . . . . . 40
        Permeation System . . . . . . . . . . . . . . . . . . . . . . . . . . . . . . . . . 41
General Membrane Operations and Maintenance Procedures . . . . . . . . 41
    Clean-In-Place System . . . . . . . . . . . . . . . . . . . . . . . . . . . . . . . . . . 43
    Monitoring Systems . . . . . . . . . . . . . . . . . . . . . . . . . . . . . . . . . . . . 44
    Typical Instrumentation and Controls . . . . . . . . . . . . . . . . . . . . . . 48
    Transmembrane Pressure Alarms . . . . . . . . . . . . . . . . . . . . . . . . . . 48
    Backwash . . . . . . . . . . . . . . . . . . . . . . . . . . . . . . . . . . . . . . . . . . . . 49
    Posttreatment Systems . . . . . . . . . . . . . . . . . . . . . . . . . . . . . . . . . . 53

## Introduction

This chapter will give a broad overview of the types of membranes used in wastewater treatment as well as general operational, maintenance, and monitoring considerations. Details that pertain to a specific membrane application will be covered in later chapters.

## Membrane Operational Terminology

Before we can discuss and compare various membrane systems, it is critical to review a few basic operational concepts for membrane systems. Definitions of key terms are as follows:

- The *influent* or the *feed water*, also referred to as the *feed stream*, to a membrane process is the stream that is to be treated.
- The *effluent* or the *permeate*, also referred to as the *filtrate* or the *product water*, is the water that has passed through the membrane.
- The *concentrate* or the *retentate*, also referred to as the *reject*, is the waste stream that is produced by direct filtration membrane processes. (Membrane bioreactors [MBRs] do not use this terminology and refer instead to mixed liquor recycle.)
- *Flux* is the volume of water that passes through a membrane per unit of time and per unit of surface area of the membrane; it is measured in either liters per square meters per hour or gallons per day per square foot and is affected by water temperature. The flux is often normalized to a standard temperature of 25°C (77°F) to account for fluctuations in water viscosity. The flux in MBR applications is particularly affected by the mixed liquor concentration.
- *Recovery* is the percentage of feed that is converted to permeate in direct filtration applications. Membrane bioreactor systems do not refer to recovery.

- Contaminant removal is defined as the percentage of a contaminant removed from the feed stream by direct membrane-filtration processes. Contaminant removal may be calculated for any parameter of interest (turbidity, total suspended solids, total organic carbon, etc.).

- *Size exclusion* is the removal of particulate matter by sieving. In very basic terms, if a membrane has a maximum pore size of $x$, then no particles larger than $x$ could pass through.

- The *molecular weight cutoff* (MWCO) or nominal molecular weight cutoff is an alternative means of measuring which particles will or will not pass through a membrane. Molecular weight is measured in Daltons. A Dalton (Da) is a non-SI unit of mass equal to the *unified atomic mass unit* (one-twelfth of the mass of a carbon-12 atom in its nuclear and electronic ground state). It is often used in biochemistry and molecular biology and also to describe the size of contaminants in water relative to their ability to pass through membranes, although the method was never approved by the Conférence Général des Poids et Mesures (IUPAC, 1997).

- Membrane *fouling* is the reduction of the flux through a membrane caused by the buildup of contaminants. The fouling of a membrane can take place either at the surface (*macrofouling*) or inside the pore (*pore fouling* or *microfouling*). Fouling can be reversible (restored by air scouring or chemical cleaning) or nonreversible.

- The *transmembrane pressure* (TMP) is defined as the difference between the average feed/concentrate pressure and the permeate pressure. It is effectively the driving force associated with any given flux for low-pressure membranes. The TMP of the membrane system is an overall indication of the feed-pressure requirement; it is used, with the flux, to assess membrane fouling.

- *Net driving pressure* is the pressure available to drive the feed water through the membrane minus the permeate and osmotic backpressure.

- *Permeability* of a membrane is the combination of flux and TMP: the flux per unit pressure of driving force.

Additional definitions may be found in the glossary at the end of this manual.

## Membrane Classifications

Membranes are classified in general by the membrane pore size, the applied pressure, and the MWCO. The removal mechanisms of pollutants are distinctively different between the various classes of membranes, as indicated in Table 2.1. Microfiltration (MF) and ultrafiltration (UF) remove impurities (suspended solids and particles) by size exclusion (sieving). In nanofiltration (NF) and reverse osmosis (RO), the removal is achieved by diffusion and charge (electrostatic) exclusion as well as size exclusion.

Figure 2.1 compares the four types of membrane separation processes with conventional depth filtration and shows the typical pollutants removed by each.

The following observations can be made from Figure 2.1:

- Microfiltration and UF do not remove ions or, in general, dissolved solids;
- Microfiltration can remove larger bacteria and pathogenic microorganisms;
- Ultrafiltration also removes some viruses; and
- Reverse osmosis removes most solids, including dissolved salts and metal ions.

Depending on the specific contaminants of concern for a wastewater-treatment system, the correct type of membrane must be selected to achieve effluent quality goals.

For the purposes of this publication, both immersed (submerged) and pressurized MF and UF systems are referred to as low-pressure

Table 2.1 General Characteristics of Membranes (the Pore Sizes and Operating Pressures Listed are General Ranges; They May Vary Slightly from One Membrane Manufacturer to the Next) (Adapted from Stephenson et al., 2000) (Courtesy of IWA Publishing, Ltd., London)

| Membrane Operation | Driving Force | Mechanism or Separation | Molecular Weight Cutoff, Range (Da) | Pore Size, Range, Microns[a] | Operating Pressure, psi[b] |
|---|---|---|---|---|---|
| MF | Pressure or vacuum | Sieve | >100 000 | 0.1–10 | 1-30 |
| UF | Pressure | Sieve | >2000–100 000 | 0.01–0.1 | 3–80 |
| NF/low pressure RO | Pressure | Sieve+ Solution/diffusion+ exclusion | 300–1000 | 0.001–0.01 | 70–220 |
| RO | Pressure | Solution/diffusion + exclusion | 100–200 | <0.001 | 800–1200 |

[a]One micron is $4.0 \times 10^{-5}$ in.
[b]To convert to kPa, multiply psi by 6.89.

membrane systems. Comparatively, NF/RO systems are high-pressure membranes that operate at significantly higher pressures to overcome the higher inherent head loss in these systems.

# Terminology for Membrane System Components

As technology advances and applications broaden in any field, there is an evolution of the related terminology. When advancements are rapid, new terms and definitions are constantly being added. Over time, standard terms become established. The terminology for wastewater membrane treatment is not firmly established at this time. For the purposes of this publication, terms and definitions have been selected and used consistently throughout this document. However, it is important

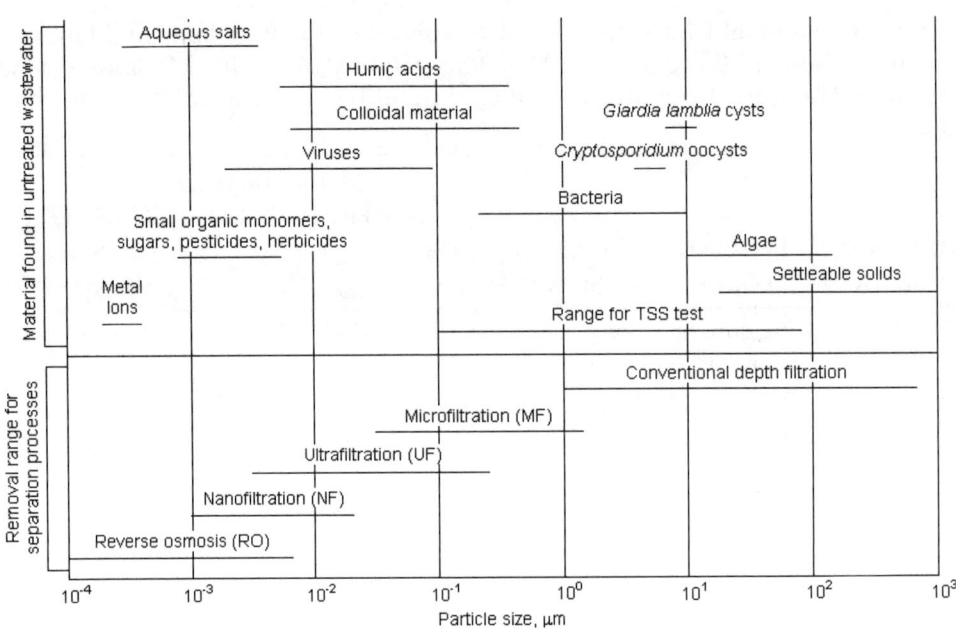

FIGURE 2.1—COMPARISON OF MEMBRANE FILTRATION PROCESSES.

to be aware that some terms, particularly those associated with membrane equipment configurations, may not be used in the same way in other references. At this point in time, the terminology used for the equipment can vary significantly from manufacturer to manufacturer.

One cause for some of the terminology differences is that membranes actually form a bridge between two fields that have traditionally been quite independent: the drinking water industry and the wastewater treatment industry. Nanofiltration/RO membranes have been used for drinking water applications for much longer than they have been used for wastewater posttreatment. Therefore, the terminology used for these systems aligns more with drinking water treatment terminology. Membrane bioreactors applications, on the other hand, are closely linked to traditional wastewater treatment. The terminology used for the MBR systems is more aligned with the wastewater treatment industry, such as mixed liquor recycle instead of retentate. The use of MF, UF, NF, or RO membranes to polish secondary effluent or posttreat tertiary effluent from wastewater treatment systems results in the use of both drinking water and wastewater treatment terminology. In this section, we will discuss membrane system component terminology to provide a common language upon which to base the rest of this publication.

The term *membrane module* is used to describe a complete unit consisting of fiber or sheet membranes; support frame or structure; and the feed inlet, retentate, and outlet permeate ports as applicable.

*Immersed membrane systems* typically include the process tank(s) with interconnecting piping for the feed water and drains, membrane units with air-scour and filtrate connections, and frames used to support membrane units. For pressurized systems, the process tanks and frames are replaced with *pressure vessels* used to contain the membranes and racks or frames on which the vessels are mounted. For cross-flow UF systems, piping and connections for the concentrate will also be provided.

In immersed systems, a series of membranes are assembled into *membrane frames* that are mounted in the process tank(s). Figure 4.5 shows a typical configuration for an immersed system. As with conventional gravity filter systems, the feed water supply and drains are hydraulically connected to tanks or cells by a manifold of pipes or common channels. Unlike gravity filters, the filtrate connections from all frames in one cell are connected to a common filtrate header that is, in turn, connected with the headers from other membrane tanks.

In contrast, each *pressure system frame* supports a group of pressure vessels that are interconnected by common feed water, filtrate, and drain blocks or piping. Feed water is pumped into the system, and the residual pressure is used to convey the filtrate to storage or a pumping station if higher pressure is required. For systems with multiple frames, the process connections from each frame are connected into larger piping manifolds. Figure 2.2 is a photograph of an individual frame of pressurized MF membranes. Figure 2.3 is a photograph of several pressure system frames.

An *array* is an assembly of cartridges in pressurized systems.

In pressurized systems, a *cartridge* is a manufactured canister of membranes with feed, permeate, and retentate connections.

FIGURE 2.2—ONE FRAME OF PRESSURIZED MF MEMBRANE MODULES (COURTESY OF USFILTER).

For immersed systems, a *cassette* (also known as a *rack*) is a removable assembly of membrane modules with connections for air and permeate.

A *module* (also sometimes referred to as an *element*) is a collection of membranes intended to be mounted and replaced as a unit.

A *train* is a combination of multiple cassettes or racks, with common permeate piping and air-scour piping that are aligned in parallel with other trains to treat a portion of a waste stream. A train is also sometimes referred to as a *pass*.

A *stage* is one portion of a train that includes membranes operating in series.

*Membrane Equipment and System Overview* ■ 23

FIGURE 2.3—SEVERAL FRAMES OF A PRESSURIZED UF SYSTEM (COURTESY OF USFILTER).

A *membrane unit* is an assembly of membranes intended to be removed from an immersed system as a unit. This term is sometimes used interchangeably with cassette or rack.

Additional definitions can be found in the glossary at the end of this publication.

# Membrane Materials and Configurations

The design and manufacture of membranes is a complex process that is beyond the scope of this publication. However, as a result of design and manufacturing processes, each membrane has specific features that affect the design and operation of membrane-related components. For example, materials used in membranes and modules can limit the options for types of chemicals used for backwash and cleaning procedures. Also, the membrane type and pore size determine the level of treatment of which the membrane system is capable. The range of available membrane materials is very diverse and varies in both chemical composition and physical structure. We will cover some of the basics in this section.

The physical structure of membranes can be described as either *microporous* or *asymmetric* (see Figure 2.4). Microporous membranes are cast from one material (they are homogenous), and they can be either uniform in pore size (*isotropic*) or varied (*anisotropic*). Isotropic membranes are also sometimes referred to as symmetric membranes. With anisotropic membranes, the surface with the smaller pore size is used as the selective or filtering surface. Integral asymmetric membranes (also known as "skinned" membranes) are cast in one process and con-

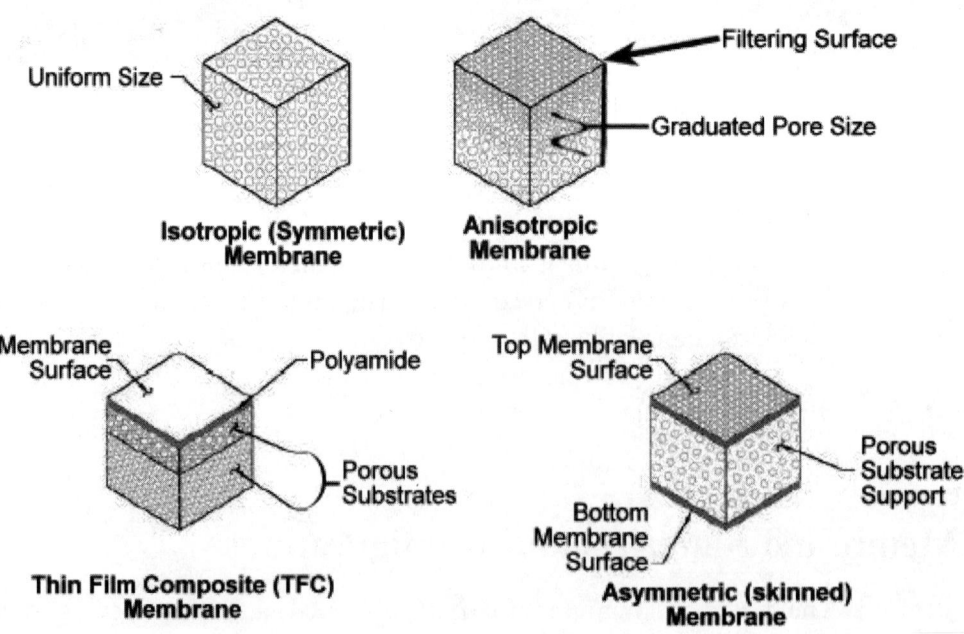

FIGURE 2.4—CROSS-SECTIONS OF VARIOUS MEMBRANE TYPES.

sist of a very thin (less than 1 μm) layer referred to as the "skin" and a thicker (up to 100 μm) porous layer that adds support and is capable of high water flux. Thin-film composite membranes are a more recent development. They are made by bonding a thin cellulose-acetate, polyamide or other acetate layer (typically 0.15 to 0.25 μm thick) to a thicker porous substrate, which provides structural stability.

The materials used to fabricate membranes can be categorized as organic and inorganic. Organic materials are either cellulose-based or consist of modified organic polymers. Inorganic materials such as metals and ceramics are used in niche industrial applications but are often cost-prohibitive in wastewater treatment. A list of available membrane materials and their characteristics is shown in Table 2.2.

There are advantages and disadvantages to the use of the various membrane materials. A summary of the advantages and disadvantages of the more common membrane materials is included in Table 2.3.

Size-exclusion membranes require a pressure differential across the membrane to force water through it. Application of a positive pressure to the feed stream "pushes" the permeate stream through the fiber. Membranes operated by applying pressure to the feed side of the membranes must be housed in a vessel capable of withstanding elevated operating pressures, commonly referred to as a pressure vessel. Size and pressure requirements of pressure vessels vary with each manufacturer. However, other size-exclusion membranes, called immersed membranes, use a negative pressure (vacuum) to "pull" the permeate stream through the fiber and do not require pressure vessels. These membranes are often in a sturdy frame or support structure to protect the individual membrane elements from damage and are placed directly into a new or existing tank.

The types of membrane elements commonly used in wastewater treatment include

- Flat sheet,
- Hollow fibers,
- Tubes,
- Spiral-wound cylinders, and
- Rotating flat plates.

Table 2.2  Membrane Material Characteristics (Adapted from Stephenson et al., 2000) (Courtesy of IWA Publishing, Ltd., London).

| Membrane | Manufacturing Procedure | Structure[a,b] | Applications |
|---|---|---|---|
| Ceramic | Pressing, sintering of fine powders | 0.1–10 μm pores | MF, gas separation, separation of isotopes |
| Stretched | Stretching of partly crystalline foil | 0.1–1 μm pores | Filtering of aggressive media, sterile filtration, medical technology |
| Etched polymers | Radiation followed by acid etching | 0.5–10 μm cylindrical pores | Analytical and medical chemistry, sterile filtration |
| Supported liquid | Formation of liquid film in inert polymer matrix | Liquid-filled porous matrix | Gas separations, carrier-mediated transport |
| Symmetric microporous | Phase inversion reaction | 0.05–5 μm pores | Sterile filtration, dialysis, membrane distillation |
| Integral asymmetric microporous | Phase inversion reaction followed by evaporation | 1–10 nm pores at membrane surface | UF, NF, gas separation, evaporation |
| Composite asymmetric microporous | Application of thin film to microporous membrane | 1–5 nm pores at membrane surface | UF, NF, gas separation, evaporation |

[a]One micrometer, μm (or micron) = $4.0 \times 10^{-5}$ in.
[b]One nanometer (nm) = $4.0 = 10^{-8}$ in.

Table 2.3  Advantages and Disadvantages of Various Membrane Materials (Adapted from Stephenson et al., 2000) (Courtesy of IWA Publishing, Ltd., London).

| MATERIAL | ADVANTAGES | DISADVANTAGES |
| --- | --- | --- |
| Cellulose acetate | Inexpensive and easy to fabricate<br>Solvent cast | Poor thermal stability, not recommended for use at temperatures higher than 30°C (86°F)<br>Poor chemical tolerance, limited to a pH range of 3 to 6<br>Poor mechanical stability, highly biodegradable<br>Limited chlorine resistance |
| Polyamide | Good thermal stability, can be used at temperatures above 50°C (122°F)<br>Good chemical stability, wide operating pH range of 3 to 11<br>Greater permeability than cellulose acetate membranes | Sensitive to chlorine |
| Polypropylene | Can withstand moderately high temperatures | Less resistant to chemicals than polytetrafluoroethylene membranes<br>Stretched fiber—elongated pores<br>Sensitive to chlorine |
| Polysulfone | Good thermal stability, can be used at temperatures up to 75°C (167°F)<br>Good chemical stability, tolerant to a wide range of pH values (1 to 13)<br>Fairly good resistance to chlorine<br>Easy to fabricate<br>Good chemical resistance to aliphatic hydrocarbons, fully-halogenated hydrocarbons, alcohols and acids | Poor chemical resistance to aromatic hydrocarbons, ketones, ethers, and esters<br>Relatively low pressure limits —690 kN/m$^2$ (100 psig) for flat sheet and 172 kN/m$^2$ (25 psig) for hollow fiber membranes |

TABLE 2.3 Advantages and Disadvantages of Various Membrane Materials (Adapted from Stephenson et. al., 2000) (Courtesy of IWA Publishing, Ltd., London) *(continued)*.

| MATERIAL | ADVANTAGES | DISADVANTAGES |
| --- | --- | --- |
| Polytetrafluoroethylene | Very hydrophobic<br>Excellent organic resistance<br>Excellent chemical stability to strong acids, alkalis, and solvents<br>Wide operating temperature range, −100°C to 260°C (−148°F to 500°F) | Only available in MF pore sizes<br>Expensive |
| Polyvinylidene fluoride | Autoclavable<br>Good resistance to solvents | Less chemically resistant to chemicals than polytetrafluoroethylene membranes<br>Only available in MF and UF pore sizes |
| Titanium dioxide | Good thermal resistance | Very expensive |
| Zirconium dioxide | Good chemical resistance<br>Good mechanical resistance | Brittle materials<br>Limited to MF and UF pore sizes |

Pleated-cartridge filters can also meet the U.S. Environmental Protection Agency's definition as a component of a membrane filtration process. They are commonly used in wastewater treatment, mostly as prefilters or to concentrate viruses from treated wastewater. These membranes are generally designed as disposable units and will not be discussed further in this publication.

Flat-sheet membranes are comprised of a series of flat membrane sheets and support plates. A single cassette can house many membrane cartridges that are slid into grooves for support. Each membrane cartridge is composed of a support plate with sheets of membrane material welded to both sides, as shown in Figure 2.5. Mixed liquor is filtered as it flows between and parallel to the cartridges, which are typically spaced approximately 10 mm (0.4 in.) apart. Flat-sheet membranes are used in immersed systems in MBRs and for effluent filtration.

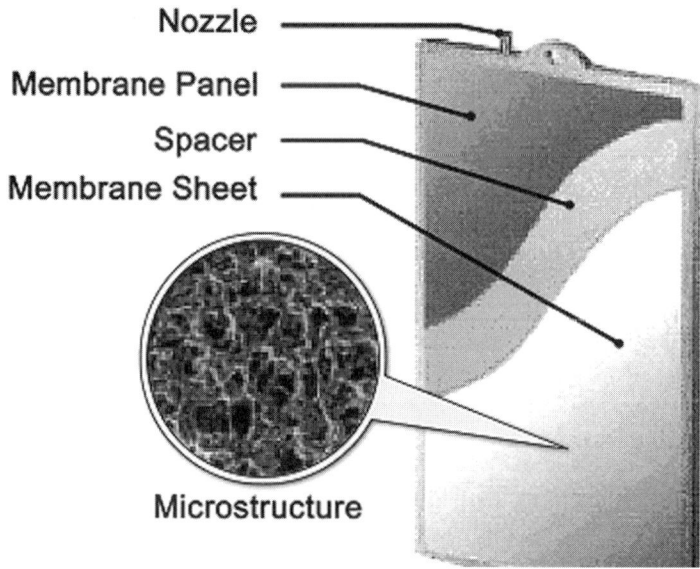

FIGURE 2.5—CROSS-SECTION VIEW OF A FLAT PLATE AND FRAME MEMBRANE (COURTESY OF KUBOTA CORPORATION).

Hollow-fiber membranes are used in both MBR and effluent filtration applications. The hollow-fiber membrane module consists of a bundle of hundreds to thousands of hollow-fiber membranes. Typical membrane cassettes/racks for MBR applications consist of the hollow fibers mounted on a frame with the permeate extraction from either or both ends of the membrane. The membrane cassettes/racks that are immersed in the process tank are installed either horizontally or vertically and are operated in an "outside to in" flow mode. An illustration of an outside-in, hollow-fiber membrane is shown in Figure 2.6, and a photo of a hollow-fiber membrane rack for an immersed system is shown in Figure 2.7.

Hollow-fiber membranes that are used for the filtration of secondary effluents can either be immersed or housed in pressure vessels. The keep the feed water and product streams isolated. The vessels are designed to prevent leaks and pressure losses and are constructed of a wide variety of materials, such as plastic, reinforced fiberglass, steel, and stainless steel. Figures 2.8 and 2.9 show cutaway views of

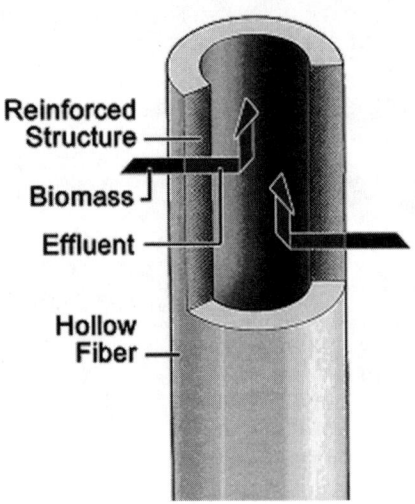

FIGURE 2.6—CROSS-SECTION VIEW OF AN OUTSIDE-IN FLOW HOLLOW-FIBER MEMBRANE (COURTESY OF ZENON ENVIRONMENTAL, INC.).

FIGURE 2.7—PHOTOGRAPH OF A SUBMERGED HOLLOW-FIBER MEMBRANE CASSETTE (COURTESY OF ZENON ENVIRONMENTAL, INC.).

FIGURE 2.8—CUTAWAY OF AN ENCLOSED HOLLOW-FIBER UF MODULE (COURTESY OF GENERAL ELECTRIC).

hollow-fiber membrane pressure vessels. The feed can be applied to the inside of the fiber (inside-out flow) or to the outside of the fiber (outside-in flow).

In tubular systems, the membranes are cast on the inside of a support tube and then placed into a pressure vessel. The feed water is pumped through the feed tube, and the product water is collected on the outside of the tube while the concentrate continues to flow through it. A tubular membrane configuration is shown in Figure 2.10.

While hollow-fiber and tubular membranes are both hollow tubes, they differ by one or more orders of magnitude in size. The outside diameter of hollow-fiber membranes is often on the order of 1 to 2 mm (0.03 to 0.06 in.) while the outside diameter of tubular membranes is often on the order of 25 mm (1 in.). The smaller diameter of the hollow-fiber

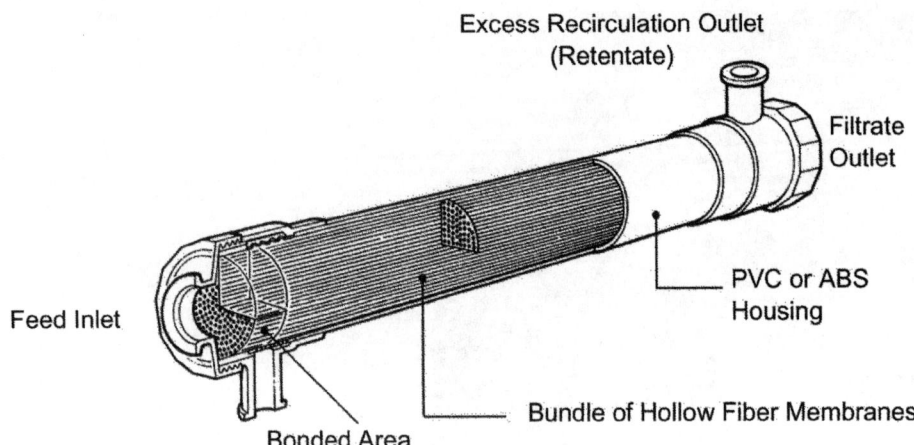

FIGURE 2.9—CUTAWAY SCHEMATIC OF AN INDIVIDUAL ENCLOSED HOLLOW-FIBER MEMBRANE MODULE (COURTESY OF PALL CORPORATION).

membranes allows for hundreds and, in some instances, thousands of these fibers to be grouped together in one bundle. These bundles of hollow-fiber membranes have a very high packing density (available surface area per unit volume), which is not possible with tubular membranes. High packing densities and low pressure requirements associated with hollow-fiber membranes tend to result in low life-cycle cost, which has caused this configuration to emerge as the most popular configuration for large municipal facilities.

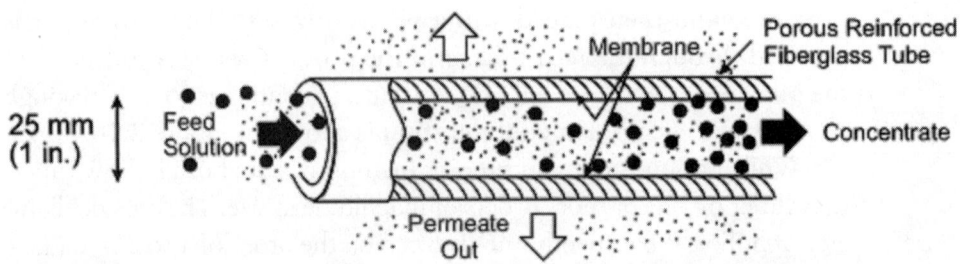

FIGURE 2.10—TUBULAR MEMBRANE CONFIGURATION (COURTESY OF KOCH MEMBRANE SYSTEMS).

*Membrane Equipment and System Overview*     ■     33

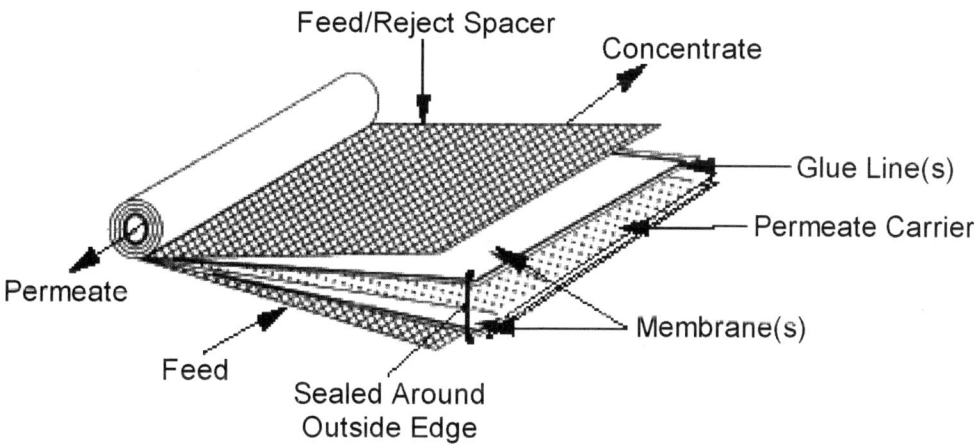

FIGURE 2.11—COMPONENTS OF SPIRAL-WOUND MEMBRANE.

In the spiral-wound membrane, a flexible permeate spacer is placed between two flat membrane sheets as illustrated in Figure 2.11. The membranes are sealed on three sides, and the open side is attached to a perforated pipe. Flow through the system is outside-in. A flexible feed spacer through the system follows a spiral flow pattern, as shown in Figure 2.12.

Individual membrane modules are typically placed in series in a pressure vessel. Spiral-wound membrane modules (or elements) are

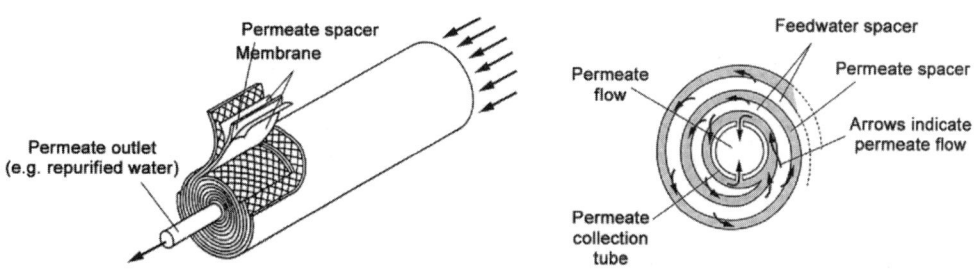

FIGURE 2.12—SPIRAL-WOUND MEMBRANE CONFIGURATION (ADAPTED FROM CRITES AND TCHOBANOGLOUS, 1998; COURTESY OF THE MCGRAW-HILL COMPANIES).

typically 200 mm (8 in.) in diameter and 1 m (40 in.) long. However, several membrane manufacturers are researching and producing prototype membranes that are longer—up to 1.5 m (60 in.)—and wider—300, 400, and 450 mm, respectively (12, 16, and 18 in., respectively). Each membrane pressure vessel is designed to hold several 1-m-long (40-in.-long) elements. Microfiltration/UF pressure vessels typically hold four elements, and RO pressure vessels typically hold six to seven elements. Figure 2.13 shows a spiral-wound pressure vessel.

Rotating flat-plate membranes are sometimes used in immersed MBR configurations. This design is intended to help alleviate fouling issues. Flat-sheet membranes and flow spacers are mounted to support disks that are connected to a rotating axis. The axis spins the membranes, creating shear forces that help to clean the membranes and reduce fouling. Table 2.4 provides a comparative summary of the characteristics of MF and UF membranes from several different manufacturers. This provides an idea of the variety of different MF and UF membrane products available today.

## Membrane System Configurations and Components

Standard membrane system configurations consist of the membranes themselves and the pretreatment and posttreatment systems, which all depend on the source water and effluent quality requirements. Mem-

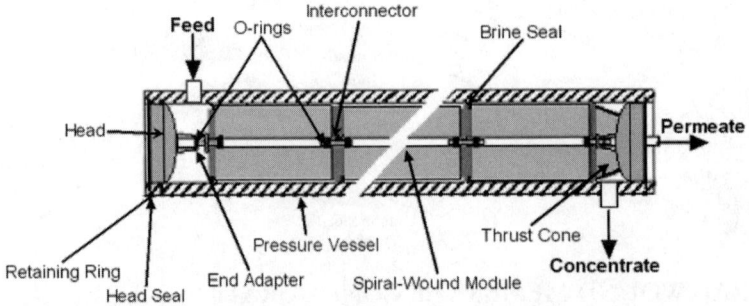

**FIGURE 2.13—TYPICAL PRESSURE VESSEL FOR SPIRAL-WOUND MEMBRANES.**

Table 2.4 Summary of Key Features of Selected MF and UF Membranes.

| Feature | Mfg. A | Mfg. B | Mfg. C | Mfg. D | Mfg. E | Mfg. F |
|---|---|---|---|---|---|---|
| Membrane material[a] | Hydrophilic PES | Hydrophilic PVP and PES | PVDF | PP | PVDF | PP | PVDF | Composite PVDF on nylon | PVDF on nylon |
| Flow configuration | Inside-out | Inside-out | Outside-in | Outside-in | Outside-in | Outside-in | Outside-in |
| Pressure configuration | Pressure | Pressure | Pressure | Pressure | Immersed | Immersed | Immersed | Immersed |
| Type | Hollow fiber UF | Hollow fiber UF | Hollow fiber MF | Hollow fiber MF | Hollow fiber MF | Hollow fiber MF | Hollow fiber UF | Hollow fiber UF |
| Cleaning chemicals | NaOH, NaOCl, citric acid ($C_6H_8O_7$), HCl, $H_2SO_4$ | NaOCl, $H_2O_2$, NaOH, EDTA, HCl, citric acid | NaOCl, citric acid, NaOH | Citric acid, $H_3PO_4$, HCl, NaOH, detergent solutions | Citric acid, $H_3PO_4$, HCl, NaOCl | Citric acid, $H_3PO_4$, HCl, NaOH, detergent solutions | Citric acid, $H_3PO_4$, HCl, NaOCl | NaOCl, citric acid based solution | NaOCl, citric acid based solution |
| Chlorine tolerance (ppm)[b] | 2 [continuous feed]; 100 [maximum], 200 000 (ppm-hr) [over lifetime] | 500 [maximum]; 250 000 [over lifetime] | 1000 [during cleaning] | 0 [not chlorine tolerant] | 200 [during cleaning] | 0 [not chlorine tolerant] | 200 [during cleaning] | 2000 [during cleaning] | 2000 [during cleaning] |
| Fiber inside diameter (mm)[c] | 0.8 and 1.2 | 0.8 and 1.5 | 0.7 | 0.25 | 0.39 | 0.5 | 0.5 | 0.72 | 0.35 |
| Fiber outside diameter (mm) | 1.3 and 2.0 | | 1.3 | 0.55 | 0.65 | 0.8 | 0.8 | 1.95 | 0.70 |

Table 2.4  Summary of Key Features of Selected MF and UF Membranes (continued).

| Feature | Mfg. A | Mfg. B | Mfg. C | Mfg. D | Mfg. E | Mfg. E | Mfg. F |
|---|---|---|---|---|---|---|---|
| Pore size ($\mu m$)[d] | 150 kDa (MWCO)[f] | 150–200 kDa (MWCO) | 0.1 | 0.2 | 0.1 | 0.2 | 0.1 | 0.04 | 0.02 |
| Maximum TMP (psi)[e] | 20 | 18 | 30 | 30 | 30 | 12.5 | 12.5 | 12 | 10 |
| Typical TMP (psi) | 3–22 | <7 | 5–25 | 5–25 | 5–25 | 3–8 | 3–8 | 3–8 | 3–8 |
| Backwash procedure | Water w/ NaOCl or $H_2O_2$ | Water w/ NaOCl, HCl or $H_2O_2$ | Air and water w/ NaOCl | Air and water | Air and water | Air and water | Air and water | Water w/ NaOCl | Water w/ NaOCl |

[a]PES = polyethersulfone; PVDF = polyvinylidene fluoride; PP = polypropylene; PVP = polyvinylpyrrolidone.
[b]1 ppm = 1 mg/L.
[c]1 mm = 0.04 in.
[d]1 micrometer (micron) = $4.0 \times 10^{-5}$ in.
[e]1 psi = 6.89 kPa.
[f]MWCO = molecular weight cutoff, expressed in kDa.

brane bioreactors are a specific subset, with the membranes installed within the activated sludge process. In addition, every membrane system requires a cleaning system and a monitoring system. Although details regarding these systems for specific applications are included in the following chapters, we will address components that are common to all membrane systems here.

## PRETREATMENT REQUIREMENTS

Membranes are a critical and costly component of the treatment process, and additional costs associated with improved pretreatment will typically be recovered over the life of the plant in terms of extending membrane life and improved long-term system performance.

## FLOW EQUALIZATION

Determining the need for and size of flow equalization is a critical decision, as it can affect the overall sizing and operation of membrane equipment. A membrane filtration system ultimately functions as the limiting hydraulic flow restriction in a wastewater treatment plant. A properly designed membrane system is designed to take varying flow conditions and redundancy into account; however, there will always be a finite limit to the flux and, therefore, to the flow that a membrane system can process. Each of the membrane equipment manufacturers uses its own sizing criteria. Criteria include the duration and magnitude of a variety of peak-flow design conditions in addition to the expected degree of fouling and the nominal average daily flow for MBR and effluent polishing systems. When designing for reuse applications, the water requirements of users will typically be the deciding factor for system sizing.

## FINE SCREENING

Fine screening is particularly important for membranes because the accumulation of trash and fibrous materials such as hair and paper found in municipal wastewater can hinder membrane performance and, ultimately, diminish membrane life. Figure 2.14 illustrates an automatic screen. The accumulation of trash on the membrane surface

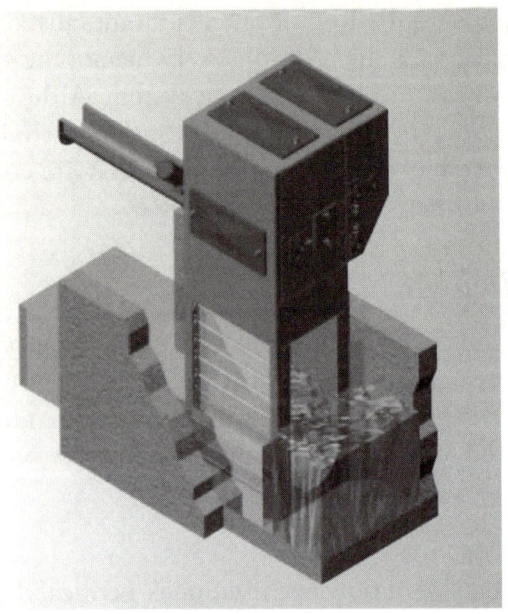

**FIGURE 2.14—ILLUSTRATION OF AN AUTOMATIC SCREEN (COURTESY OF EIMCO [JONES AND ATWOOD, LTD.]).**

can effectively reduce the membrane surface area available for permeation, thus increasing the TMP required to move the same volume of permeate through the system. Trash and other fibrous materials that accumulate on the membrane surface will also result in increased tension on the membranes themselves, potentially damaging the membranes. Attempting to physically remove trash to clean the membranes to restore permeability may cause more harm than good because membranes may be accidentally pulled loose and/or broken. Although this is particularly true for hollow-fiber membranes, flat-plate membrane systems are not immune to the problems of improperly screened influent, as trash can still accumulate between the plates. Additionally, any membrane system using coarse-bubble aeration as a means of scouring the membrane system can be affected by insufficiently screened wastewater. Fibrous material can potentially block the holes in the coarse-bubble diffusers and result in improperly scoured membranes and a reduction in treatment capacity.

Each membrane manufacturer has different guidelines for screening. Most state a maximum opening of 2 mm (0.08 in.), whereas others allow up to 3 mm (0.12 in.). In practice, most new facilities are being designed with screens with openings smaller than 2 mm (0.08 in.), and the trend is for finer screening. Membrane plants with screen openings of smaller than 1 mm (0.04 in.) have recently been designed and commissioned. Screens with openings in the 2- to 3-mm (0.08- to 0.12-in.) range can be controlled to allow for additional matting, reducing the effective screen opening even further.

Two preferred features of a screen for a membrane wastewater-treatment system are bypass prevention and two-dimensional or omnidirectional openings.

The best protection that can be offered to the membranes is to include the finest screening possible at the head of the plant while minimizing the potential for bypass around the screen mechanism.

Effluent polishing and advanced posttreatment membrane systems typically include additional fine screens, strainers, or cartridge filters on the treatment plant effluent line immediately before the membranes to protect them from algae or debris that has blown into the process tanks beyond the headworks screens.

# GRIT REMOVAL

Grit can damage membranes by abrasion, especially in the turbulent environment created by the air scour. In larger plants, grit will settle in process tanks long before it ever reaches the membranes. Some means of grit removal is recommended for smaller plants, as smaller process tanks have a greater potential for carryover. Membrane plants often have some sort of equalization tank upstream that will generally be sufficient to settle out any grit. Grit removal is typically included in large-scale municipal wastewater facilities to protect the membranes and other equipment such as pumps.

# FATS, OILS, AND GREASE REMOVAL

Typically, fats, oils, and grease (FOG) is not an issue with membranes that are used for tertiary filtration or MBR treatment of municipal wastewater. Grease could cause significant fouling problems if it were to reach the membrane system; however, because the wastewater has undergone a biological treatment process before reaching the membranes, the FOG has already been biodegraded into simpler substances that will not cause fouling issues with the membranes.

## MEMBRANE SYSTEM EQUIPMENT COMPONENTS

Components used in typical MF and UF treatment facilities include membranes and associated hardware; a permeation system, which might include a vacuum priming system; a flux-maintenance system, such as air scour and chemical cleaning; an integrity monitoring system; and a programmable logic controller (PLC) and associated instrumentation. Microfiltration and UF membranes are manufactured in either an immersed or an encased (pressurized) configuration. Both configurations have similar components; however, significant variations exist in the components used by each manufacturer. More information is provided below.

### Membranes and Associated Hardware

The organization of multiple membrane units into larger groupings is a function of the overall membrane system design. Because of the relatively large number of membrane units installed in even relatively small municipal systems, loss of one unit has little effect on system capacity; however, the loss of major subgroupings can seriously reduce production capacity. Redundancy and the enhanced reliability that it provides can be achieved by installing redundant groups or subgroups of membranes or by having uninstalled spares kept on site. The overall system design determines the organization of pressure vessels and membrane units, whereas both influent flow and/or filtrate use can determine the level of redundancy required for the membrane treatment facility. A low-pressure system treating only a portion of the secondary effluent from a larger upstream plant for reuse via urban irrigation may require no redundancy because loss of filtrate for short periods of time will cause little damage. However, an MBR system treating plant influent or an MF system providing pretreatment for an RO system supplying industrial process water for a critical application may not be able to cease operations for more than a few minutes or hours.

Permeation System

All low-pressure membrane systems require a pressure gradient across the membranes. Most low-pressure membrane installations use a pump to provide the pressure differential needed to force water through the membrane; although, in some situations, the differential might be provided by gravity. Immersed membrane systems typically use an effluent-side vacuum permeation system to "pull" filtrate across the membrane, whereas pressurized membrane systems require feed pumps to "push" the water through the membrane. When the driving force is a vacuum, it is limited by the vapor pressure of water; whereas, with pressurized systems, the driving force is limited by the membrane's physical strength, which is often several times that of the vapor pressure. Whether vacuum or pressurized, the permeation system can be configured as either a dedicated or a common (manifold) system. Dedicated permeation systems have one pump per membrane sub-grouping and manifold permeation systems use a common header to hydraulically join all membrane groups to one set of pumps.

# General Membrane Operations and Maintenance Procedures

Operating strategies for membrane systems are selected to maintain the desired operating flux. These strategies differ, depending on the type of membrane system.

Backwashing of the membrane is a common technique used with low-pressure, hollow-fiber membranes to maintain the design operating flux of the system. During each backwashing sequence, normal filtration is stopped for a group of membranes, and filtrate is pumped across the membrane in the reverse direction by a backwashing pump. Small systems may use the same pump for filtrate and backwash pumping. Backwashing. allows for the removal of material retained on the feed side of the membrane, which helps maintain the design operating flux. Material, or cake, retained on the feed side of the membrane may also be removed by backwashing with high-pressure air or draining and backwashing. The air–water–membrane interface can impart significant force on solids to remove them from the membrane surface because of the surface tension of air and water.

The number of backwash systems installed at a membrane facility depends on the size of the facility, organization of the membranes, and process layout. Although installation of only one backwash system is common at small- to medium-sized membrane facilities where each membrane train is backwashed independently of the others, large systems may require multiple systems.

In some systems, particularly MBRs, a relax cycle is used instead of backwashing. Relaxation of membranes is provided by simply stopping filtration for a short period of time. Experience with some membranes has shown relaxation to be as effective as backwashing in maintaining flux.

Maintaining the flux of an immersed membrane system is slightly different because membranes are not confined in pressure vessels. Immersed systems rely on combinations of air scour, backwashing with filtrate, and relaxation. Air scour is an important component of all immersed systems and is used to generate turbulence along the surface of the membranes. Coarse-bubble aeration is used to introduce the scour air from orifices or diffusers located in the bottom of the membrane frame or immediately under the frame. As the bubbles rise from the bottom of the frames, they impart turbulence to the feed stream, which helps dislodge contaminants retained on the feed side of the membrane. Most immersed systems provide scour air continuously, but air can also be applied on a time cycle or only during backwash. Some immersed systems drain the tank contents after each backwash to remove accumulated debris, whereas others drain the tank contents intermittently or infrequently. Pressure-vessel systems may also blow down the vessels intermittently.

Chemicals are used to enhance the backwashing procedure; chemical metering pumps are used to inject chemicals to the filtrate used for backwashing. Chemicals for backwashing are typically injected to a separate backwashing tank or directly to the backwash piping. Selection of a chemical or combination of chemicals for chemically enhanced backwashes (CEBs) depends on the nature of the foulants in the feed water and the type of membrane. Sodium hypochlorite, hydrochloric acid, and hydrogen peroxide are some of the chemicals that have been used for CEBs; although, for secondary effluent applications, hypochlorite is probably the most common choice. Membrane compatibility must be verified with the manufacturer before the injection of any chemical to the membrane system, and all chemicals should be well mixed before they come in contact with the membranes.

# CLEAN-IN-PLACE SYSTEM

When the transmembrane pressure cannot be restored to normal operating values by backwashing, chemical cleaning of the membranes is required to maintain the design operating flux of the system. Chemical cleaning is typically required at intervals of months on low-fouling feed water and at intervals of weeks on high-fouling feed water. As with backwashing, the state of the art for chemical cleaning is constantly changing as new membranes are developed and more experience is gained from operating full-scale systems. Early low-pressure membrane facilities had to physically remove the membrane units from the process tanks and place them in a dedicated cleaning tank. Current designs use multiple membrane tanks or cells so that a group of membranes can be removed from service and cleaned in place. The term "clean-in-place" or "CIP" is often used to describe the systems required to provide this function.

One CIP system is typically sufficient for most facilities, but multiple units may be provided depending on the size of the facility, organization of the membranes, and process layout. Typically, each membrane block is cleaned in place independently of the others. The CIP systems on large systems should include two mixing tanks with associated pumps, mixers, and heaters so that two different chemicals can be prepared at the same time. A cartridge filter should be provided on the discharge line to the membranes to remove any particulate matter that is present in the bulk chemical or added during solution preparation. Piping for the cleaning solution should take suction from the solution tanks, convey solution to the membrane tanks, and allow return to the solution tank or drainage. On large systems, appropriate considerations must be given to storage and handling of bulk cleaning chemicals. Figure 2.15 is a photograph of a chemical cleaning system at a membrane facility.

## MONITORING SYSTEMS

Depending on the effluent permit requirements or reuse application, continuous or periodic monitoring of the integrity of the membranes is conducted to verify filtrate quality and detect equipment problems. Integrity testing can be conducted by either direct or indirect methods. Direct tests measure changes in pressure, air flow, or sound whose magnitude is a

**FIGURE 2.15—CHEMICAL CLEANING SYSTEM AT A LOW-PRESSURE MEMBRANE FACILITY.**

direct function of breaches in the membrane system. Indirect methods rely on measurements of water-quality parameters such as turbidity or particle counts in the filtrate. Declines in filtrate water quality are a surrogate measure of defects in membrane system integrity.

Many direct integrity tests are based on bubble-point theory. Once a microporous membrane has been wetted, water is held in the pores by surface tension, and relatively high air pressure is required to push water out of the pores. The required air pressure is inversely proportional to the size of the opening. When air pressure is applied to a wet membrane and the pressure is gradually increased, a pressure will be reached at which air will begin to flow through the largest pore (or defect). This is the bubble point. Common direct methods based on bubble-point theory include the diffusive airflow test, the pressure decay (PDT) or pressure hold test, the acoustic sensor and audible tests, and the bubble-point test. These methods test membrane integrity by introducing pressurized air to the membrane system and monitoring specific testing conditions. The diffusive airflow test measures increased airflow through the membranes resulting from a defect, while the PDT test measures changes in pressure. Visual inspection for air bubbles in the discharge piping of individual pressure vessels or water surrounding immersed membranes during pressure testing assists in locating individual leaks. No specialized instrumentation is required for these membrane integrity tests beyond accurate pressure or flow measurement. Both the acoustic sensor test and the audible test rely on detecting the sound of air passing through a defect, and a hydrophonic sensor and a stethoscope are required to conduct these membrane integrity tests.

Pressure (or vacuum) decay testing is the most common direct method in current use with low-pressure membranes used to treat secondary effluent, and many MF and UF systems have automated membrane-integrity testing and monitoring systems. Typically, water will be drained from the filtrate side of the membranes; the filtrate side will be isolated; and air pressure, typically approximately 100 $kN/m^2$ (15 psi), will be applied. Pressure loss or air flow can then be used to evaluate membrane integrity.

Advantages and disadvantages for MF/UF membrane integrity tests are listed in Table 2.5.

Additional information regarding direct and indirect methods of membrane-integrity testing can be found in the references listed in Appendix II.

Table 2.5 Summary of Integrity Monitoring Methods.

| Integrity Test Method | Advantages | Disadvantages |
|---|---|---|
| *Direct Methods* | | |
| Pressure decay test | ■ Can monitor entire rack of membranes simultaneously<br>■ Ability to detect single fiber breaks and small holes<br>■ Standard on most membrane systems and highly automated<br>■ Can test membranes and downstream plumbing for leaks<br>■ Ability to maintain aseptic quality of system if applied to filtrate side | ■ Need for increasingly sensitive pressure transducers when test is applied to large numbers of modules<br>■ Potential to yield false-positive results if the membrane is not fully wetted<br>■ Limitation in log removal equivalency (sensitivity) |
| Diffusive airflow test | ■ Similar advantages as the pressure decay test; typically not included as standard equipment<br>■ More sensitive than the pressure hold test as currently applied<br>■ Ease of conducting and accuracy of test (when measuring water displacement) | ■ Sensitivity to temperature (viscosity)<br>■ Limited full-scale applications to verify performance<br>■ Manual application as currently applied, though the test can be automated<br>■ Not standard equipment for most MF/UF systems |
| Bubble point test | ■ Ease of conducting and interpreting results<br>■ Ability to identify specific compromised fibers and leaking seals | ■ Labor intensive for large plants—manual application<br>■ Only able to pinpoint leaks identified by other test methods<br>■ Only practical as a diagnostic test for the repair of individual modules |

Table 2.5 Summary of Integrity Monitoring Methods *(continued)*.

| Integrity Test Method | Advantages | Disadvantages |
|---|---|---|
| Sonic sensing analysis | <ul><li>Identifies compromised module and general location of integrity breach</li><li>Easy to use</li><li>Potential to be developed into a continuous, online monitoring method</li></ul> | <ul><li>Manual application</li><li>Limited use in the water industry to date</li><li>Labor intensive for large plants</li><li>Subjective interpretation of results</li><li>Not practical for submersed systems</li></ul> |
| Particle counting | <ul><li>Continuous monitoring of filtrate water quality</li><li>Sensitive to minor changes in water quality</li><li>Widespread use and familiarity in water industry</li></ul> | <ul><li>Difficult to calibrate</li><li>Imprecision between different (even well-calibrated) instruments</li><li>Susceptible to sensor clogging</li><li>Susceptible to counting entrained air as particles, or particles from growth in instrument tubing</li><li>Relatively high cost</li></ul> |
| Particle monitoring | <ul><li>Continuous monitoring of filtrate</li><li>Significantly lower cost than particle counters</li><li>No calibration required</li><li>More sensitive to integrity breaches than turbidimeters, but less sensitive than particle counters</li></ul> | <ul><li>Infrequent use in water industry</li><li>Less sensitive than particle counters</li><li>Susceptible to sensor clogging</li><li>Susceptible to entrained air as particles, or particles from microbial growth in instrument tubing</li><li>Provides only a relative index of particle concentration</li></ul> |
| Turbidity monitoring | <ul><li>Continuous monitoring of filtrate</li><li>Widely used in water industry and for MBR applications</li><li>Significantly lower cost than particle counters</li></ul> | <ul><li>Except for MBR applications, relative insensitivity to breaches in membrane integrity compared with other methods</li><li>Susceptible to counting entrained air as particles</li></ul> |

## TYPICAL INSTRUMENTATION AND CONTROLS

Instrumentation installed in any process revolves around parameters of interest that are specific to the process, and low-pressure membrane facilities are no different. Typical parameters monitored in MF and UF treatment facilities filtering secondary effluent include flow (volume), pressure, and turbidity or particle counts. Other parameters that may also be of interest include nutrients, fecal or total coliforms, total dissolved solids/conductivity, UV-254 absorption (as a surrogate for organic content), and iron concentration. Inclusion of the necessary instrumentation to monitor these parameters is not necessary at most MF and UF treatment facilities.

Configuration of the control system will be primarily established by the equipment manufacturer because the system is integral to the successful operation of the membrane process. Automation is made possible by the use of PLCs. The PLCs control virtually every aspect of operation for a membrane treatment facility, including operating flux, backwashing sequences, membrane-integrity tests, and chemical addition. Modifications of the basic control system are possible, depending on the specific goals and constraints of a particular system. To protect membranes from potential catastrophic damage, automatic to manual flow control, automatic high-flow shutdown control, and automatic high-pressure shutdown control systems are recommended (at a minimum). Should downstream process be sensitive to increases in particulate matter, an additional automatic high-turbidity or particle-count shutdown may be desired.

## TRANSMEMBRANE PRESSURE ALARMS

Transmembrane pressure alarms automatically monitor the TMP for a group of membranes. An alarm is activated if the differential pressure across the membranes exceeds a predetermined setpoint, which is typically specified by the membrane manufacturer. Monitoring TMP requires accurate pressure indicators and transmitters immediately upstream and downstream of the membranes. Differences in elevation between the pressure gauges and the membranes must be taken into account in calculating the TMP. Feed water in immersed systems is open to the atmosphere so that a pressure (vacuum) measurement is only required on the filtrate side of the membranes. The number of TMP alarm systems installed at a membrane facility depends on the size of the facility and organization of the membranes. Typically, one TMP alarm system is installed on each membrane skid.

## BACKWASH

The backwash system provides for the automatic initiation of a backwashing sequence and controls backwash pumps and blowers (where applicable); feed or filtrate pumps; and associated automatic valves on the feed, filtrate, and drain lines. A backwashing sequence is typically initiated at a predetermined time interval (typically between 10 and 30 minutes); however, backwashing sequences may also be initiated at a predetermined TMP increase. Provisions for the manual initiation of a backwashing sequence are also typically provided. Should the membrane facility inject chemicals to enhance the backwashing sequence, the operation of chemical-metering pump(s) is also controlled by the backwash system. Backwash flux, cross-flow velocity, chemical addition (where applicable), sequence, duration, and transmembrane pressure are all important variables in backwashing low-pressure membranes. Variation of one or more of these parameters is often required before finding the optimum backwash protocol for an individual system.

During a high-pressure air backwashing procedure, normal filtration is stopped for a group of membranes, and water may be drained from the filtrate or feed side of the membrane fibers by opening a drain valve. Low-pressure air may be applied to remove all water. Once water has been removed from the fibers, the drain valve is closed, and the lumens are pressurized with high-pressure air (approximately 620 $kN/m^2$ [approximately 90 psi]). After the pressure setpoint is reached, backwash valves on the feed side of the membranes are opened, allowing the air to expand through the walls of the membrane fibers to dislodge contaminants from the feed side of the membrane. Dislodged contaminants are then removed by opening feed and backwash valves to allow the feed side of the fibers to be flushed with feed water using a cross-flow pattern. Finally, air is removed from both sides of the membranes and the system is returned to the normal filtration mode. A complete air backwash cycle will require approximately 2.5 minutes and is repeated at intervals of approximately 20 minutes.

Pressurized membrane systems operating with an inside-out flow configuration (the reference being to the normal direction of filtration) when backwashing rely on pumping filtrate from the outside of the fiber sequentially to the lumen of the fiber, from the filtrate side to the feed side, to physically dislodge accumulated particulate matter. Such a backwash sequence for inside-out, hollow-fiber systems may start with a forward flush in which feed water is forced out the concentrate line to remove any particulate matter in the lumens. Alternatively, the feed side of the membrane may be drained by blowing down the feed piping to empty the vessel and remove some particulate matter. After the forward flush or drain, backwash water is then pumped through the membrane from the filtrate side into the lumens and discharged from one end of the fibers, followed by discharge from the other end of the fibers. If a CEB is being used, the fibers are soaked in the chemical solution by stopping the backwash pumps for a fixed period, and the entire system is backflushed before resuming normal operation. Optimization is necessary to balance the chemical costs and flux maintenance benefits. The entire sequence may require less than a minute but can last longer and is repeated every 15 to 60 minutes. Chemicals may be added to every backwash but are typically only used a few times per day.

As with pressurized systems, backwashing protocols for immersed membrane systems are different for each manufacturer. Immersed systems will also operate in the normal filtration mode until the TMP setpoint for backwashing is exceeded or the desired time interval has passed. Backwashing will begin in some systems by stopping feed to the membrane cell while continuing filtration until the tank water level drops to the top of the membranes. When the draining is complete, filtration will stop and air scouring of the membranes will begin. Figure 2.16 shows membranes in an immersed cell being scoured with air. As the air scour agitates the membrane surface, filtrate will be pumped back through the membranes from the filtrate side into the

FIGURE 2.16—IMMERSED LOW-PRESSURE MEMBRANE DURING BACKWASH.

tank. After the air scour and backwash are complete, the tank contents might be completely drained, the tank refilled with feed water, and normal filtration restarted. Backwash times for immersed membranes are similar to those used for pressurized membranes, with most sequences taking approximately 3 minutes.

Several manufacturers now rely on membrane relaxation as the primary means of maintaining TMPs. For these systems, backwashing will only be initiated if relaxation is not effective. A typical relaxation period might last approximately 10 seconds following a normal filtration period of 10 minutes.

Chemical cleaning of membranes is done less frequently than backwashing and is typically initiated manually by the operators. Chemical cleaning is required when backwashing no longer reduces the operating TMP below the maximum recommended operating value. Operation of the tank feed and drain valves, backwash pumps, blowers, and chemical-metering pump(s) used to inject chemicals to the system are controlled by the CIP system.

Membrane manufacturers typically recommend specific cleaning procedures, and any variances from factory procedures should be made in consultation with the manufacturer. The CIP process typically includes periods of membrane immersion in the cleaning solution, followed by the recirculation of the cleaning solution through the membrane. Typically, a group of membranes is offline 30 to 60 minutes during each CIP sequence; however, soaking or recirculation may be required for extended periods (e.g., 4 to 12 hours).

Standard solutions of acids, bases, and detergents can be used to clean most membranes. Depending on the type of foulant and the degree of fouling, different chemicals and procedures will be used. Chlorine in the form of sodium hypochlorite is the most common cleaning chemical used to remove biological foulants accumulated on the membrane surface. Citric acid ($C_6H_8O_7$) is a common acid used to clean membranes and is effective in removing iron fouling and mineral scales ($CaCO_3$, $CaSO_4$, and $MgSO_4$) resulting from hard water and increases in pH caused by the release of carbon dioxide from secondary effluent. Caustic (NaOH) is a commonly used base that can be effective in removing biological fouling when chlorine cannot be used. Periodic use of a different disinfectant from that used in CEBs can be helpful if bacteria are present that are resistant to the CEB disinfectant. Other chemicals used to clean low-pressure membranes include hydrogen peroxide ($H_2O_2$), hydrochloric acid (HCl), and ethylenediaminetetraacetic acid (EDTA). While cleaning is typically done with cleaning solutions at ambient temperature, better results can sometimes be obtained by heating the solution, typically to approximately 40° C (104° F) before use. Chemicals are typically mixed and heated, if necessary, in a separate CIP tank before being pumped to the membranes. Cleaning chemicals can also be directly injected to process piping; however, thorough mixing of the cleaning chemicals is required before they come in contact with the membranes. Membrane manufacturers should be consulted before injecting any chemical to the membrane system for compatibility, including recommended concentrations, pH values, and temperatures.

## POSTTREATMENT SYSTEMS

A membrane system typically incorporates posttreatment processes. The processes needed are determined by the use requirements for the permeate. In general, membrane posttreatment might include disinfection, which can be achieved through the use of gas or liquid chlorine, ozone, UV radiation, or other means.

Typical posttreatment processes for NF/RO systems often include

- Degasification, which is typically used for the removal of hydrogen sulfide ($H_2S$) and carbon dioxide ($CO_2$) from the permeate flow stream and
- Stabilization by adding chemicals into the permeate stream; typically sodium hydroxide (NaOH) is added to increase the pH, and corrosion inhibitors are added to minimize the corrosivity of the permeate water.

More detailed information on operating criteria and equipment for various membrane processes used for wastewater treatment is included in Chapters 3, 4, and 5 of this publication.

CHAPTER THREE

# Membrane Bioreactors

Introduction . . . . . . . . . . . . . . . . . . . . . . . . . . . . . . . . . . . . . . . . . . . . . . . . . . . . 58
    Applications . . . . . . . . . . . . . . . . . . . . . . . . . . . . . . . . . . . . . . . . . . . . . . . 59
    Water Quality . . . . . . . . . . . . . . . . . . . . . . . . . . . . . . . . . . . . . . . . . . . . . . 62
        Influent Quality . . . . . . . . . . . . . . . . . . . . . . . . . . . . . . . . . . . . . . . . 62
        Effluent Quality . . . . . . . . . . . . . . . . . . . . . . . . . . . . . . . . . . . . . . . 63
    Treatment Mechanism . . . . . . . . . . . . . . . . . . . . . . . . . . . . . . . . . . . . . . 64
    Equipment Manufacturers . . . . . . . . . . . . . . . . . . . . . . . . . . . . . . . . . . . 65
Process Configurations . . . . . . . . . . . . . . . . . . . . . . . . . . . . . . . . . . . . . . . . . 66
    Pretreatment Requirements . . . . . . . . . . . . . . . . . . . . . . . . . . . . . . . . . . 66
        Fine Screening . . . . . . . . . . . . . . . . . . . . . . . . . . . . . . . . . . . . . . . . . 66
        Primary Clarification . . . . . . . . . . . . . . . . . . . . . . . . . . . . . . . . . . . . 67
    Biological Treatment . . . . . . . . . . . . . . . . . . . . . . . . . . . . . . . . . . . . . . . 67
        Recycle of Mixed Liquor . . . . . . . . . . . . . . . . . . . . . . . . . . . . . . . . . 70
        Solids Retention Time . . . . . . . . . . . . . . . . . . . . . . . . . . . . . . . . . . . 71
        Quality of Activated Sludge . . . . . . . . . . . . . . . . . . . . . . . . . . . . . . 72
        Mixed Liquor Suspended Solids Concentration . . . . . . . . . . . . . . . . 72
        Oxygen Transfer . . . . . . . . . . . . . . . . . . . . . . . . . . . . . . . . . . . . . . . 73
    Membrane Separation . . . . . . . . . . . . . . . . . . . . . . . . . . . . . . . . . . . . . . 73
    Cleaning and Chemical Feed Systems . . . . . . . . . . . . . . . . . . . . . . . . . . 74
    Posttreatment . . . . . . . . . . . . . . . . . . . . . . . . . . . . . . . . . . . . . . . . . . . . 75
    Residuals Disposal . . . . . . . . . . . . . . . . . . . . . . . . . . . . . . . . . . . . . . . . 75

Equipment Configurations . . . . . . . . . . . . . . . . . . . . . . . . . . . . . . . . . . . . . 76
   Membrane Bioreactor Facility Components . . . . . . . . . . . . . . . . . . . . . . . . . . 76
   Bioreactor Equipment . . . . . . . . . . . . . . . . . . . . . . . . . . . . . . . . . . . . . . . . . . 76
      Biological Process Blowers . . . . . . . . . . . . . . . . . . . . . . . . . . . . . . . . . . . 76
      Biological Process Aeration . . . . . . . . . . . . . . . . . . . . . . . . . . . . . . . . . . 77
      Mixed Liquor Recirculation Pumps . . . . . . . . . . . . . . . . . . . . . . . . . . . 78
      Mixers . . . . . . . . . . . . . . . . . . . . . . . . . . . . . . . . . . . . . . . . . . . . . . . . . . . 80
   Membrane Filtration Equipment . . . . . . . . . . . . . . . . . . . . . . . . . . . . . . . . . 80
      Membrane Configurations . . . . . . . . . . . . . . . . . . . . . . . . . . . . . . . . . . . 80
      Permeation System . . . . . . . . . . . . . . . . . . . . . . . . . . . . . . . . . . . . . . . . 81
      Backpulse Pumps . . . . . . . . . . . . . . . . . . . . . . . . . . . . . . . . . . . . . . . . . . 83
      Backpulse Tanks . . . . . . . . . . . . . . . . . . . . . . . . . . . . . . . . . . . . . . . . . . 83
      Cleaning System . . . . . . . . . . . . . . . . . . . . . . . . . . . . . . . . . . . . . . . . . . 84
      Membrane Air Scour Blowers . . . . . . . . . . . . . . . . . . . . . . . . . . . . . . . . 85
      Drain Pumps . . . . . . . . . . . . . . . . . . . . . . . . . . . . . . . . . . . . . . . . . . . . . 86
      Air Compressors for Pneumatic System Components . . . . . . . . . . . . . . . . . . . 87
      Staging Tank . . . . . . . . . . . . . . . . . . . . . . . . . . . . . . . . . . . . . . . . . . . . . 87
   Instrumentation and Controls . . . . . . . . . . . . . . . . . . . . . . . . . . . . . . . . . . . 87
      Programmable Logic Controllers . . . . . . . . . . . . . . . . . . . . . . . . . . . . . 87
      Turbidimeters . . . . . . . . . . . . . . . . . . . . . . . . . . . . . . . . . . . . . . . . . . . . 88
      Dissolved Oxygen Meters . . . . . . . . . . . . . . . . . . . . . . . . . . . . . . . . . . . 88
Operation . . . . . . . . . . . . . . . . . . . . . . . . . . . . . . . . . . . . . . . . . . . . . . . . . . . 89
   Pretreatment . . . . . . . . . . . . . . . . . . . . . . . . . . . . . . . . . . . . . . . . . . . . . . . . 89
   Biological Process Operation . . . . . . . . . . . . . . . . . . . . . . . . . . . . . . . . . . . 89
      Sludge Wasting . . . . . . . . . . . . . . . . . . . . . . . . . . . . . . . . . . . . . . . . . . . 90
      Dissolved Oxygen . . . . . . . . . . . . . . . . . . . . . . . . . . . . . . . . . . . . . . . . . 91
      Alkalinity and pH . . . . . . . . . . . . . . . . . . . . . . . . . . . . . . . . . . . . . . . . . 91
   Membrane Filtration—Modes of Operation . . . . . . . . . . . . . . . . . . . . . . . . 92
      Permeation . . . . . . . . . . . . . . . . . . . . . . . . . . . . . . . . . . . . . . . . . . . . . . 92
      Relax . . . . . . . . . . . . . . . . . . . . . . . . . . . . . . . . . . . . . . . . . . . . . . . . . . . 92
      Backpulse/Backwash . . . . . . . . . . . . . . . . . . . . . . . . . . . . . . . . . . . . . . . 93

    Chemically Enhanced Backpulse/Backwash (Maintenance Clean) . . . . . . . . . 93
    Recovery Cleaning . . . . . . . . . . . . . . . . . . . . . . . . . . . . . . . . . . . . 94
    Flow Control . . . . . . . . . . . . . . . . . . . . . . . . . . . . . . . . . . . . . . . . 94
  Membrane Aeration . . . . . . . . . . . . . . . . . . . . . . . . . . . . . . . . . . . . . . 95
  Alarms and Trends . . . . . . . . . . . . . . . . . . . . . . . . . . . . . . . . . . . . . . 95
  Operational Changes as a Result of Events and Seasonal Changes . . . . . . . . . . 96
  Capacity Management . . . . . . . . . . . . . . . . . . . . . . . . . . . . . . . . . . . 96
  Optimization . . . . . . . . . . . . . . . . . . . . . . . . . . . . . . . . . . . . . . . . . . 96

Routine Monitoring . . . . . . . . . . . . . . . . . . . . . . . . . . . . . . . . . . . . . . . . 97
  Pretreatment . . . . . . . . . . . . . . . . . . . . . . . . . . . . . . . . . . . . . . . . . . 97
  Biological Treatment . . . . . . . . . . . . . . . . . . . . . . . . . . . . . . . . . . . . . 97
  Membrane Systems . . . . . . . . . . . . . . . . . . . . . . . . . . . . . . . . . . . . . . 99
    System Performance . . . . . . . . . . . . . . . . . . . . . . . . . . . . . . . . . . . 99
    Equipment and Instrumentation . . . . . . . . . . . . . . . . . . . . . . . . . . . 100

Maintenance . . . . . . . . . . . . . . . . . . . . . . . . . . . . . . . . . . . . . . . . . . . . 101
  Cleaning . . . . . . . . . . . . . . . . . . . . . . . . . . . . . . . . . . . . . . . . . . . . 101
    Maintenance Cleaning . . . . . . . . . . . . . . . . . . . . . . . . . . . . . . . . . 101
    Recovery Cleaning . . . . . . . . . . . . . . . . . . . . . . . . . . . . . . . . . . . 103
    Physical Cleaning . . . . . . . . . . . . . . . . . . . . . . . . . . . . . . . . . . . . 106
  Membrane Integrity . . . . . . . . . . . . . . . . . . . . . . . . . . . . . . . . . . . . . 106
    Integrity Testing . . . . . . . . . . . . . . . . . . . . . . . . . . . . . . . . . . . . . 106
    Identifying and Repairing Damaged Membrane Elements . . . . . . . . . . . . 107
  Instrument Calibration . . . . . . . . . . . . . . . . . . . . . . . . . . . . . . . . . . . 108

Troubleshooting . . . . . . . . . . . . . . . . . . . . . . . . . . . . . . . . . . . . . . . . . 109

# Introduction

A membrane bioreactor (MBR) is a combination of suspended growth-activated sludge biological treatment and membrane filtration equipment performing the critical solids/liquid separation function that is traditionally accomplished using secondary clarifiers. Low-pressure membranes (either microfiltration [MF] or ultrafiltration [UF]) are commonly used in MBRs.

Traditional treatment systems use an aeration tank, secondary clarification, and possibly tertiary filters (Figure 3.1). There are two general types of membrane systems that can be used in MBRs: pressure-driven, in-pipe cartridge systems that are located external to the bioreactor and vacuum-driven, immersed systems that are designed for installation within the bioreactor (Figures 3.2 and 3.3). Immersed membrane technologies using hollow-fiber or flat-sheet membranes are the most popular for MBR applications because they operate at lower operating pressures (or vacuums), can more readily accommodate the variations in the types of solids found in activated sludge bioreactors, and typically provide a lower life-cycle cost, particularly for municipal-scale facilities. Pressure-driven systems are more prevalent in industrial systems where waste characteristics, such as high temperatures, require the use of ceramic membranes. In its simplest form, an immersed membrane bioreactor system can combine the functions of an activated sludge aeration system, secondary clarifiers, and tertiary filtration in a single tank.

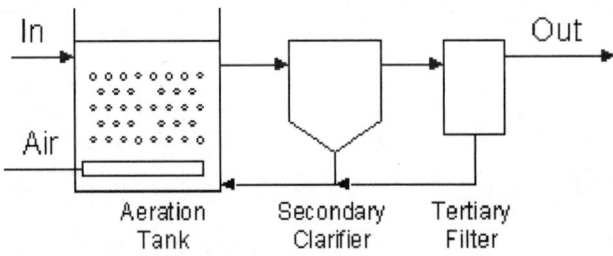

**FIGURE 3.1—CONVENTIONAL ACTIVATED SLUDGE SCHEMATIC.**

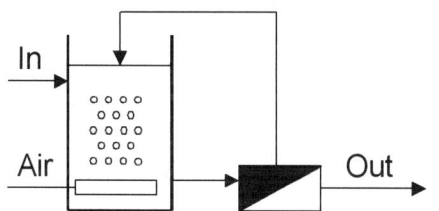

**FIGURE 3.2—EXTERNAL MBR SCHEMATIC.**

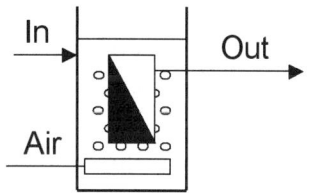

**FIGURE 3.3—IMMERSED MBR SCHEMATIC.**

## APPLICATIONS

Membrane bioreactors are effectively used to treat both municipal and industrial wastewaters. The potential benefits of MBR systems for any particular application are numerous, and some of these are listed below:

- *Exceptional effluent quality*
  - Biomass is completely retained, resulting in consistently high-quality final effluent; effluent solids concentrations are <1 mg/L (<1 ppm).
  - Compared to conventional activated sludge (CAS) plants with clarifiers, the effluent quality is less dependent on the mixed liquor suspended solids (MLSS) concentration and sludge settling properties (sludge volume index).
- *Small footprint*
  - Secondary clarifiers and effluent filters can be eliminated, thereby reducing plant footprint area.
  - Because elevated mixed liquor concentrations are possible, the aeration basin volume can be reduced, further reducing plant footprint.
  - Long solids retention times (SRTs) can be achieved in a small footprint; result is sludge production equivalent to an extended aeration system on less land than an equivalent CAS plant.

- *Modular system*
  - Membrane systems are modular in nature, allowing for ease of expansion, and flexible in configuration, making them a popular option for plants looking to retrofit older technology.
- *Robust and reliable operation*
  - System can operate within a wide range of SRTs, resulting in increased flexibility and more options for system optimization.
  - System is robust enough to handle elevated MLSS concentrations for short periods of time, allowing for flexible solids wasting schedules.
  - Processes are easily automated; operator requirements are reduced, considering that operators are not required to closely manage sludge settling issues.
- *Reduced downstream disinfection requirements*
  - A physical mechanism to remove pathogens is provided.
  - Low-turbidity effluent reduces downstream disinfection requirements; high transmissivity means less energy required for UV disinfection; effluent has minimal chlorine demand, so less is required to achieve the target residual concentration.

Membrane bioreactors often become viable treatment options for facilities requiring high-quality effluent water for reuse or discharge to sensitive receiving waters and for facilities with significant land area restrictions (both new plants and retrofits).

Membrane bioreactors can offer attractive treatment options for ski, golf, and other resort communities that are not connected to a municipal sewer system and have a particularly high demand for irrigation water. Membrane bioreactors provide resorts with the ability to treat wastewater on site in a compact facility and reclaim the water for non-potable reuse.

Sewer "mining", or "scalping", plants represent another potential application of MBR technology for water reclamation. In rapidly expanding suburban areas, potential users of reclaimed water are often not located near the main wastewater treatment plant, and installing distribution systems to convey the reclaimed water to the reuse sites is often difficult or expensive. By locating remote MBR facilities near reclaimed water users, these problems can be avoided. The "satellite" or "scalping" plants can extract or "mine" wastewater from interceptor sewers and deliver treated effluent directly to users.

Although the advantages of MBR systems are numerous, they are not suitable for every wastewater treatment application. Some potential disadvantages of MBRs include

- *Limited flow capacity*
    - Flows above the design capacity in a CAS plant typically result in incomplete treatment; however, in a membrane plant there is a hydraulic limitation to how much water the membranes can permeate. If flow requirements are greater than design or membranes are fouled during a high flow period, quantities of water beyond the system's current capacity will need to be diverted to another location for treatment or contained in a holding tank. All MBR system tanks should be designed with additional freeboard to be used as holding volume in emergency situations.
    - In light of hydraulic limitations inherent in any membrane system, particular attention must be paid to redundancy and availability of spare parts for all system components.
    - Limited peaking ability—typically, MBRs are designed with a peaking factor of 2.0 to 2.5, and any peak flows beyond this must be equalized either in an upstream holding tank or within the freeboard volume of the bioreactor tanks.
- *Limited availability of long-term data*
    - As MBRs are a relatively new technology, there is a limited amount of data available to verify long-term performance.
    - Because there are limited data available, it is impossible to verify various manufacturers' claims regarding membrane life expectancy.

- *Increased potential for foam*
  - Operating conditions in a MBR system often favor the formation of foam. Newer MBR facilities are taking this into account during the design stage and include foam management options such as surface wasting to prevent foam from accumulating.
- *Costs*
  - Although the capital cost of membranes has fallen significantly over the past five years, an MBR is an advanced treatment process and, as such, will be more expensive than a CAS plant. For plants that have selected MBR technology, however, the benefits of improved effluent quality and reduced footprint requirements are often found to outweigh the increase in capital cost.
  - Proper care must be taken to optimize the chemical usage for membrane cleaning to limit the effect of purchasing chemicals on operating costs.
  - Membrane bioreactor plants consume more energy than CAS plants. Significant consumers of energy include air scour blowers, biological process blowers, and recycle pumps.
- *System monitoring and maintenance*
  - Although highly automated and often remotely controlled, MBRs must be closely monitored to detect changes in flux rates and permeability before they escalate. Maintaining a proactive cleaning schedule can help avoid emergency situations in the future.

## WATER QUALITY

### Influent Quality

The influent to a MBR can be any wastewater stream that is treatable using an activated sludge process. Influent wastewater quality can vary significantly with geographical location and composition (proportion of domestic/industrial wastewater). Although pretreatment for the removal of grit and screenable materials is critical for the operation and maintenance of membrane systems, in many respects, the influent quality is not as important to the MBR as the MLSS concentration and the SRT of the activated sludge. These latter parameters define the quality of the material to which the membranes will be exposed and within which the membranes will be expected to operate.

Effluent Quality

The effluent from the MBR process is essentially free of solids and macrocolloidal material, as can be seen in Figure 3.4, where the mixed liquor is compared to permeated water. More importantly, the MBR facility can be designed to remove nutrients from wastewater, similar to conventional biological nutrient removal facilities with minor modifications.

In general, the effluent quality from MBR facilities is less than 1 mg/L in total suspended solids, less than 5 mg/L (5 ppm) 5-day carbonaceous biochemical oxygen demand, and less than 0.2 nephelometric turbidity units (NTUs), and it contains low levels of bacteria and viruses. As the membranes used in MBRs have pore sizes in the MF/UF range (see Chapter 1), MBRs alone will remove only minimal amounts of dissolved solids. Additionally, the membranes themselves will have no effect on

FIGURE 3.4—MIXED LIQUOR AND PERMEATED WATER.

pH or alkalinity, and some form of pH adjustment is often included in the MBR design, particularly when coagulants are used for phosphorus removal.

The effluent from MBR facilities can be used for discharge to sensitive areas, reused on public access sites, or further treated by nanofiltration or reverse osmosis (RO). Table 3.1 summarizes the typical effluent quality produced from a municipal MBR facility that is designed to achieve nutrient removal.

## TREATMENT MECHANISM

Membrane bioreactors are primarily a biological treatment process, using the activated sludge process to remove soluble and particulate matter from the wastewater to be treated. As in any activated sludge process, the key to proper operation is the successful separation of biological solids from mixed liquor to produce the desired effluent, while creating the return activated sludge that is essential for continuous operation of the process.

Table 3.1 Typical Municipal MBR Effluent Quality.

| Parameter[a] | Units | Values |
|---|---|---|
| $CBOD_5$ | mg/L[b] | <5 |
| TSS | mg/L | <1 |
| Ammonia | mg/L as N | <1 |
| Total nitrogen (with preanoxic zone) | mg/L | <10 |
| Total nitrogen (with preanoxic and postanoxic zones) | mg/L | <3 |
| Total phosphorus (with chemical addition) | mg/L | <0.2 (typical) <0.05 (achievable) |
| Total phosphorus (with Bio-P removal) | mg/L | <0.5 |
| Turbidity | NTU | <0.2 |
| Bacteria | log removal | Up to 6 log (99.9999%) |
| Viruses | log removal | Up to 3 log (99.9%) |

[a] $CBOD_5$ = five-day carbonaceous biochemical oxygen demand and TSS = total suspended solids.
[b] 1 mg/L = 1 ppm.

Unlike clarifier-based activated sludge processes, MBRs use membranes for the separation of biological solids from mixed liquor. The membrane pore sizes are minute, typically smaller than the pore sizes of filter papers used for laboratory analysis, so the separation of solids from liquids is essentially complete, and all biological solids are retained in the process for use as return sludge or for wasting.

Although membrane pore sizes are minute, in the MF or UF range, they are not so small as to be able to capture soluble organic compounds, metals, or trace contaminants such as pharmaceutical and personal care products, priority pollutants, or endocrine-disrupting compounds. Although the biological process of a MBR may adsorb or reduce such contaminants, the MBR filtration mechanism is not adequate to directly filter these materials from wastewater.

# EQUIPMENT MANUFACTURERS

There are several manufacturers that produce membrane equipment for use in MBRs. There are also several firms representing specific manufacturers and offering package MBR systems that include biological process design and membrane equipment. Most offer a choice of purchasing the equipment only or of purchasing a package that includes equipment plus process design responsibility.

It is important to understand the distinction between membrane equipment and MBR systems. Membrane bioreactor systems are biological processes that use membranes for the separation of mixed liquor solids from the water that will be discharged. Membrane equipment manufacturers have often provided the biological process design with the equipment, primarily in response to clients who desired a single-entity responsibility for these relatively new process systems. Now that more is known about MBR systems, the biological process can be designed and specified and the membrane equipment can be separately obtained from one of the competing manufacturers. As the diversity of applications has increased, the required biological process designs have become more varied and complex, and the ability of any membrane manufacturer to support the required range of plant designs has become increasingly difficult. At the same time, differences among available membrane systems can affect the process design, and membrane suppliers may define operational requirements that can affect biological processes. When systems are designed by consultants and the equipment is bid, it is absolutely critical that proper coordination occur between the designer and the successful equipment bidder.

The first MBRs were smaller than 2.6 m$^3$/min (1.0 mgd) and were a package-type design, with the manufacturer providing process design and equipment. Recently, MBRs have been commissioned with capacities of 26 to 40 m$^3$/min (10 to 15 mgd), and there are plans for 80- to 130-m$^3$/min (30- to 50-mgd) MBR plants in the near future. As MBR technology is applied to larger installations, process responsibility is typically separated from the equipment supply responsibility. There are many reasons for this, primarily the complexity of the overall plant design and the fundamental differences between process and equipment design in an optimal facility.

## Process Configurations

This section will focus on the configuration and operation of the biological process that, in combination with membrane equipment, creates the system known as a MBR. The Equipment Configurations section will focus on the specific equipment associated with both the MBR biological process and membrane equipment.

### PRETREATMENT REQUIREMENTS

Basic pretreatment requirements for membrane systems were covered in Chapter 2. Please refer back to the discussion on flow equalization, screening, and grit removal. An explanation of pretreatment requirements that are unique to MBR systems is included below.

Fine Screening

As discussed in Chapter 2, it is recommended that fine screening equipment with a maximum of 1- to 3-mm (0.04- to 0.12-in.) openings be provided to protect membranes from debris and fibrous materials. Typically, these screens are installed at the headworks of a MBR plant. An alternative approach to fine headworks screening is to use a coarser headworks screen to provide protection from debris for all of the plant's equipment and then fine screen the mixed liquor as it is pumped from the bioreactor to the membrane tanks. The disadvantages of this approach are that the screens need to be much larger, as the flow into the membrane tanks is typically five times higher than the influent flow and the screens must handle solids concentrations up to 10 000 mg/L (10 000 ppm). In spite of this, smaller plants may find this approach to screening more attractive. This approach is considered to be a form of sidestream screening. The purpose of sidestream screening is to "clean up" whatever was missed in the original headworks screening. With this type of screening configuration, the sidestream screen will typically have a smaller opening than the influent screen and will be sized to handle only a portion of the flow—25 to 50% of the average flow capacity. In plants where a separate sidestream screen is not included, the recycle piping is typically designed to allow some degree of mixed liquor rescreening simply by diverting the recycled mix liquor back through the headworks screen.

Primary Clarification

Primary clarification is not specifically required for a MBR; although, just as for other activated sludge systems, the provision of primary clarifiers can reduce the total energy required for aeration and the overall volume of the bioreactor. Primary clarification provides the additional benefit of settling out some of the undesirable trash as well as skimming off scum and floatables that would otherwise be removed through fine screening. Some membrane manufacturers will require less stringent fine screening if primary clarification is included in the process.

# BIOLOGICAL TREATMENT

The biological design of MBR systems has been reported for a variety of different combinations of effluent criteria relating to ammonia, total nitrogen, and phosphorus. Design criteria are, therefore, becoming available for distinct types of treatment applications, including nitrification (Corona, California; Woodstock, Georgia; and The Hamptons, Georgia); nitrification with chemical addition for phosphorus removal (Port McNicoll, Ontario); nitrogen removal (Key Colony, Florida); nitrogen removal with chemical phosphorus removal (Arapahoe County, Colorado); and biological nutrient removal (Traverse City, Michigan, Cauley Creek, Georgia; Running Springs, California; and Bandon Dunes, Oregon). A recently completed Water Environment Research Foundation (WERF) research project is intended, in part, to provide process design and operations information to owners and operators involved with MBR systems. Figures 3.5 to 3.8 provide simplified flow schematics of a MBR system, illustrating a few of the basic possible process configurations.

In addition to these configurations, various other advanced biological processes can be combined or incorporated to MBR design. Some of these include

- Nitrogen removal, taking advantage of the long SRT typically made possible by a MBR and including an anoxic zone and recycle flow for the denitrification of the nitrate produced. The recycle of mixed liquor from the membrane area directly to an upstream anoxic zone can be considered, using the Modified Ludzack-Ettinger (MLE) process (Figure 3.6); however, the high dissolved oxygen concentration that is often found in the recycle can reduce denitrification efficiency. A recycle of the mixed liquor to the upstream aerobic zone, combined with the recycle of mixed liquor from just beforer the membrane zone to the anoxic zone, may be preferred for MBR operations. In this way, the oxygen concentration in the stream being recycled to the anoxic zone may be lower and can be

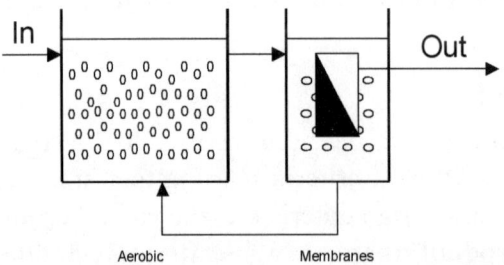

FIGURE 3.5—MEMBRANE BIOREACTOR—NITRIFICATION ONLY.

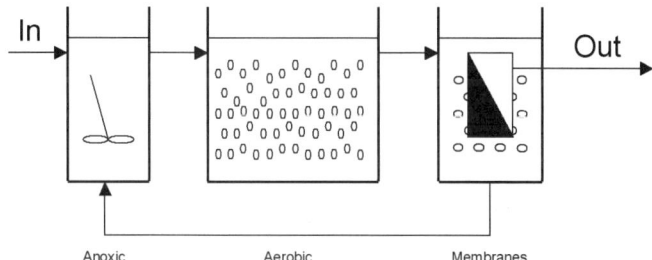

**FIGURE 3.6—MEMBRANE BIOREACTOR TOTAL NITROGEN REMOVAL.**

controlled by the operator in that portion of the aerobic zone. An additional benefit of a dual recycle configuration is the ability to completely decouple the solids recycling requirements from the denitrification requirements. The downside to a dual recycle configuration is the higher capital and operational cost associated with two sets of recycle pumps instead of one. An alternative approach for reducing the dissolved oxygen concentration in the recycle stream is to design a small deaeration zone upstream of the anoxic zone.

- Supplemental addition of an external carbon source such as methanol to a post-anoxic zone to further enhance denitrification can be effective, particularly in plants required to reduce total nitrogen to less than 3 mg/L (3 ppm).

- The addition of chemicals for phosphorus removal can be practiced with MBRs (Figure 3.7) in a similar manner as for conventional activated sludge processes. Two observations have been made with MBRs:

  ➢ Because virtually all of the particulate phosphorus is removed in a MBR, the metal salt dosage required to achieve a certain treatment objective may be lower with MBRs.

  ➢ The addition of metal salts will have a beneficial effect on membrane flux, as increasing the size of the flocs makes for more easily filtered mixed liquor and reduces membrane fouling.

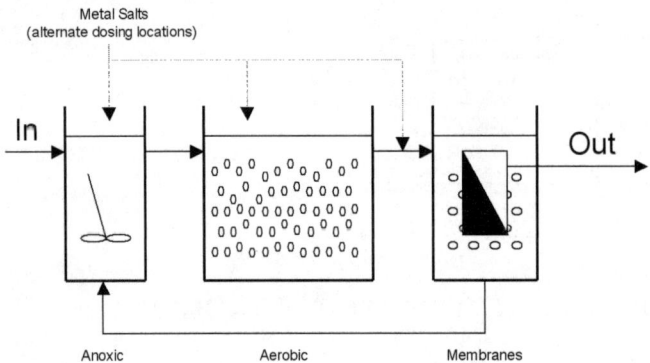

**FIGURE 3.7—MEMBRANE BIOREACTOR—NITROGEN REMOVAL WITH CHEMICAL PHOSPHORUS REMOVAL.**

- Biological nutrient removal can be achieved using many of the typical processes proven to support the growth of polyphosphate-accumulating organisms. In this case, the preservation of soluble organic material is more important than for nitrogen removal, and the unintended recycle of dissolved oxygen to the anoxic zone must be avoided. The operator would also need to monitor and avoid the transfer of nitrate from the anoxic zone to the anaerobic zone, and for this control the operator should be provided with the ability to adjust the various recycle flows. Many alternate configurations exist for achieving biological phosphorus removal. One possible configuration is shown in Figure 3.8.

Recycle of Mixed Liquor

Just as in any conventional activated sludge process, some sludge must be recycled from the solids/liquid separation device back to the front of the biological process to "activate" the biological process. In the case of MBRs, the recycle can be 200 to 400% of the plant flow, and a minimum recycle is required to flush the membrane area and control the concentration of mixed liquor (MLSS) in the area of the membranes. If the recycle rate is too low, the MLSS in the membrane tank will escalate rapidly, making operation unsustainable. The main objective of the high recycle rate is to redistribute the sludge inven-

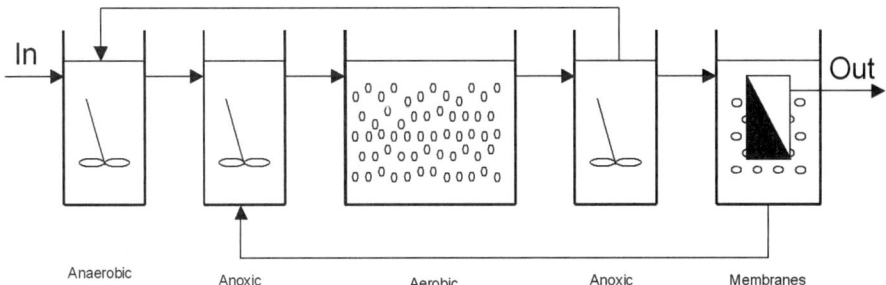

FIGURE 3.8—MEMBRANE BIOREACTOR—BIOLOGICAL NUTRIENT REMOVAL.

tory and minimize membrane fouling associated with elevated MLSS concentrations. By returning this same recycle stream to a preanoxic zone, denitrification can be readily achieved.

It is important to note that the recycle of mixed liquor from the membrane area can contain very high concentrations of dissolved oxygen, on the order of 2 to 6 mg/L (2 to 6 ppm) instead of the virtual absence of dissolved oxygen in the return sludge from a clarifier. This oxygen cannot be significantly controlled because the air flow provided by the air scour blowers must be significant enough to provide a minimum shearing action across the surface of the membranes. Although this will result in a somewhat less efficient denitrification process, the majority of MBR facilities in operation today with total nitrogen requirements do operate this way (see Figure 3.6) and are able to achieve total nitrogen requirements of less than 10 mg/L. To account for the elevated dissolved oxygen in the recycle stream, anoxic zones in MBRs with single recycle streams and without dedicated deaeration zones are proportionally larger to compensate for the reduction in denitrification efficiency. Where a CAS plant may have an anoxic zone that is 15 to 20% of the total bioreactor volume, a MLE-based MBR plant with a single recycle stream may have an anoxic zone that is 20 to 40% of the total bioreactor volume.

Solids Retention Time

Most of the initial MBR systems were designed with extremely long SRTs, on the order of 30 to 70 days, and only recently have MBRs been routinely operated at fewer than approximately 20 days. One initial concern was the reduced flux resulting from short SRT operation, presumably caused by the fouling effect of extracellular excretions from younger sludges. Recent experience with immersed membranes has indicated that biopolymer fouling is not strongly related to SRT, provided that the SRT is at least long enough to perform nitrification. In these cases, fouling can be effectively controlled by automated in situ membrane-cleaning methods. The selection of SRT based on treatment requirements is, therefore, now possible, and recent pilot studies indicate that SRTs as low as 8 to 10 days are feasible. As in any activated sludge process, nitrification rates are highly temperature dependent. Whereas an SRT of 12 days may be suitable at 18 °C (64 °F), to achieve the same level of nitrification an SRT of 20 days may be required at 10 °C (50 °F).

Quality of Activated Sludge

Biological treatment systems rely on many types of microorganisms being present in the bioreactor. Bacteria have key roles, which include the conversion of soluble and particulate organic compounds into biomass and gaseous waste products, conversion of ammonia to nitrate (nitrification), and conversion of nitrate to nitrite and nitrogen gas (denitrification). Higher forms of microorganisms, such as protozoa and rotifers, play an important role in consuming particulate organics. The microbiological conditions in a MBR system are similar to those in an activated sludge system. However, there are some noted differences. Flocs in the MBR systems appear to be smaller, and this may be a result of the higher shearing action of air scour systems and mixed liquor recycle pumps. By replacing the gravity settling requirements in secondary clarifiers, MBR systems avoid issues of filamentous sludge bulking and other floc settling and clarification problems. In addition, the aeration tank MLSS is no longer controlled by the solids loading limitations that are inherent with sizing secondary clarification facilities. Typically, MBR systems operate at higher MLSS concentrations that range from 6000 to 20 000 mg/L (6000 to 20 000 ppm).

Mixed Liquor Suspended Solids Concentration

Immersed MBR systems have typically operated with MLSS concentrations between 8000 and 12 00 mg/L (8000 and 12 000 ppm), with occasional operation between 15 000 and 18 000 mg/L (15 000 and 18 000 ppm). Operating in this range has reduced the bioreactor volume required and minimized waste sludge handling and stabilization because of high waste-sludge concentration. However, high MLSS has been shown to reduce membrane flux and the aeration alpha factor, leading to higher aeration energy requirements. Current design practice is to assume the MLSS to be closer to 8000 to 10 000 mg/L (8000 to 10 000 ppm) to ensure adequate oxygen transfer and allow for higher membrane flux. Operators should carefully monitor the MLSS concentration at the membrane to ensure that it does not become excessively high and does not exceed the manufacturer's recommendation.

Oxygen Transfer

At MLSS concentrations that are higher than intended by design, the demand for oxygen can increase significantly, both because of the higher concentration of biological activity and the higher associated SRT. In some cases, the demand can exceed the volumetric capacity of typical oxygenation systems and, as a result, the operator may observe a decrease in the dissolved oxygen concentration that can be maintained in the aerobic zones. The oxygen-transfer capacity of the aeration system must also be carefully understood. Further, the high MLSS concentration itself may affect transfer efficiency by reducing the alpha factor. Immersed membranes are typically provided with shallow coarse-bubble air to agitate the membranes as a means to control fouling. This "membrane aeration" provides some oxygenation but at low efficiency. Typical dissolved oxygen concentrations in the various zones of an MBR are

- Anoxic: 0.0 to 0.5 mg/L (0.0 to 0.5 ppm)
- Aerobic: 1.5 to 3.0 mg/L (1.5 to 3.0 ppm)
- Membranes: 2.0 to 6.0 mg/L (2.0 to 6.0 ppm)

# *MEMBRANE SEPARATION*

Similar to the traditional biological suspended-growth processes that use clarifiers that are designed for the removal of suspended solids, MBRs accomplish the same removal objectives using membranes in place of the clarifiers. The primary difference is that microorganisms are more completely separated from the water using membranes, resulting in a higher-quality effluent than can be achieved with secondary clarifiers and granular media filters, neither of which are required in the MBR process.

For successful operation, membrane equipment systems for MBRs include various combinations of air-scour, backpulse, relaxation, and chemical cleaning systems to maintain performance and flux. Membranes can be operated across a range of flows, or flux, as long as an adequate differential pressure is provided across the membranes. Immersed MBR systems have limited differential pressure (less than 69 kPa [10 psi]), hence flux rates are more restricted than with external, pressure-driven MBRs. Experience with MBR membrane equipment suggests that average flux rates between 14 and 25 $L/m^2 \cdot h$ (8 and 15 gpd/sq ft) are sustainable when the MLSS is 15 000 mg/L (15 000 ppm) or less.

## CLEANING AND CHEMICAL FEED SYSTEMS

Membranes are critical to successful plant operation and must be maintained and cleaned on a regular basis to ensure that the desired system filtration capacity is provided. Various automated and event-triggered cleaning methods are discussed in detail in the Maintenance section of this chapter. Several of the recommended cleaning procedures involve the use of chemicals to remove residues from the membrane surfaces. Acid solutions are used to clean the membranes of inert deposits, whereas dilute chlorine solutions are used to eliminate organic growth and fouling of the membranes.

Cleaning chemicals can either be injected inline as the membranes are backwashed or pumped into a clean-in-place (CIP) tank—cleaning solution from the CIP tank is typically backwashed into the membranes or transferred into the membrane tank.

Other chemical feed systems that may be found in an MBR plant include

- Coagulants for phosphorus removal (e.g., ferric chloride, aluminum sulfate (alum), sodium aluminate) and
- Sodium hydroxide for pH and alkalinity control.

## POSTTREATMENT

The permeate from a MBR system will be high quality with very low turbidity on the order of 0.2 to 0.5 NTU. Disinfection of the effluent is often still required because membranes do not provide an absolute barrier to bacteria and viruses. Because of the clarity of MBR effluent, UV transmissivity will be significantly higher than that of a filtered conventional activated sludge plant effluent (approximately 75% for MBR effluent compared with approximately 65% for filtered CAS effluent). Therefore, UV disinfection can be much more economical for MBR treatment systems. Similarly, because of the very low presence of particulate matter, the effectiveness of chlorine disinfection will also be higher for MBR effluents.

High-quality effluents for reuse may require the further removal of trace contaminants, which can be achieved by adding RO treatment for the permeate stream. Refer to Chapter 5 of this publication for more information on the operation of RO systems for this application.

## RESIDUALS DISPOSAL

Most MBR applications to date have been in smaller plants with aerobic sludge stabilization followed by biosolids disposal, either by liquid land application or dewatered biosolids reuse. At larger plants, thickening the waste sludge for anaerobic or aerobic digestion and then dewatering the biosolids are more typical. To date, little research has been conducted on thickening and/or high-rate stabilization of waste MBR sludge. The research that has been conducted has not found any significant differences in sludge thickening or stabilization characteristics, except that gravity thickening is not as successful because of the already high solids concentration of the mixed liquor before thickening.

A few plants also use membranes for thickening waste activated sludge before further treatment or disposal. These membrane-thickening applications typically use the same type of membranes as are used in the MBR but operate at much lower flux rates (in the range of 3 to 8 L/m$^2$·h [2 to 5 gpd/sq ft]). These systems are capable of thickening waste activated sludge up to 4 to 5% of the total solids concentration.

# Equipment Configurations

## MEMBRANE BIOREACTOR FACILITY COMPONENTS

As described earlier in this chapter, a typical plant that includes a membrane bioreactor system consists of the following major units:

- Preliminary treatment system (headworks),
- Biological process tankage (bioreactor),
- Biological process blowers,
- Membrane filtration system,
- Air scour blowers,
- Backpulse system (manufacturer-dependent),
- Mixed liquor recirculation system,
- Cleaning system,
- Posttreatment, and
- Waste sludge treatment and disposal.

A schematic of the major components of a complete membrane bioreactor facility is shown in Figure 3.9. The equipment described in this section will be limited to the MBR process equipment, specifically the equipment used in the bioreactor and membrane filtration system.

## BIOREACTOR EQUIPMENT

The major equipment used in the bioreactor is similar to the equipment used in the bioreactor of a conventional activated sludge process. This includes process aeration blowers, biological process aeration, and mixers.

Biological Process Blowers

Depending on the size of the plant, the process aeration blowers can be either positive-displacement or centrifugal. The process blower system is typically designed as a common group of blowers (duty plus on-line standby) that provides air to all biological process trains. All of

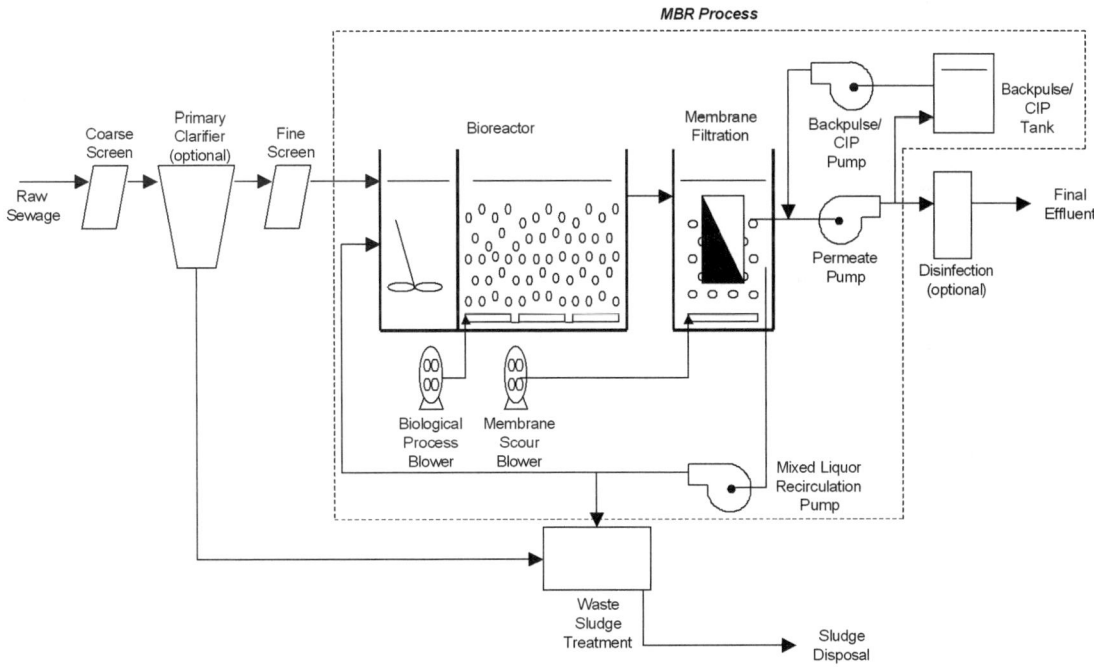

FIGURE 3.9—PROCESS FLOW DIAGRAM OF A TYPICAL MBR FACILITY.

the blowers discharge to a common air supply manifold that delivers air to the individual diffuser grids in each aerobic zone. The process aeration blowers are typically separate from the membrane air scour blowers, although the two systems may share a common standby.

Biological Process Aeration

Biological process aeration is provided by typical aeration systems used in conventional activated sludge processes, including fine-bubble aeration, coarse-bubble aeration, and jet aeration. Fine-bubble diffusers with full floor coverage are the most common type of aeration system used in the aerobic zones of bioreactors because of their higher oxygen-transfer efficiency (Figure 3.10). Tubular or disc-type membrane fine-bubble diffusers are most frequently used.

**FIGURE 3.10—FINE-BUBBLE AERATION SYSTEM.**

Mixed Liquor Recirculation Pumps

Mixed liquor recirculation in a MBR system can be designed in one of two ways:

- Pumping mixed liquor from the bioreactor to the membrane tanks and returning the mixed liquor from the membrane tanks to the bioreactor by gravity or
- Allowing gravity flow of mixed liquor from the bioreactor to the membrane tanks and pumping the mixed liquor from the membrane tank to the bioreactor.

The first approach requires pumping $(R + 1)Q$, whereas the second requires pumping $RQ$ ($R$ = the mixed liquor recirculation ratio and $Q$ = the influent flow). A number of different pump types can be used for mixed liquor recirculation—submersible or high-capacity end-suction centrifugal pumps are most common. Axial-flow pumps are well suited for this application as well because of the high-flow, low-head requirements (Figure 3.11).

FIGURE 3.11—RECIRCULATION PUMPS.

The sizing of the recirculation pumps in a membrane bioreactor is dictated by a few important factors:

- Sufficient flow to avoid buildup of the mixed liquor solids within the membrane tanks and ensure proper solids distribution between the biological process tanks and membrane tanks.
- Sufficient nitrate return flow to the unaerated zones at the head of the bioreactor to achieve the required levels of pre-denitrification and the target effluent nitrate concentration.
- Some membrane manufacturers design their facilities using a proprietary jet aeration system to scour the membranes; in these systems, recirculation pumps are used to achieve the proper mixed liquor flow rate and head through the jets at the base of the membrane modules; these systems require that the mixed liquor be pumped from the bioreactors to the membrane tanks.

For systems where the mixed liquor recirculation is used solely to dilute the mixed liquor solids within the membrane tank, the recirculation pump is typically sized for two to four times maximum day flow. If the pumps are also used as part of a denitrification system, they may be sized as large as three to six times maximum day flow. Some systems use the recirculation pumps as part of the two-phase jet system, which combines fluid transfer with air scour energy. Total dynamic head is based on head losses through the pump and piping systems or through the jets where applicable.

### Mixers

The unaerated zones in the bioreactor are typically equipped with a dedicated mixer in each zone to keep the solids in suspension. Unaerated zones may include deaeration (or deoxygenation) zones, preanoxic and postanoxic zones, and anaerobic zones. While submersible mixers are most common, surface mixers can also be used. The mixers are used to ensure adequate mixing within each zone for proper contact between biomass, substrate, and energy source (nitrate or oxygen) and to prevent short-circuiting across any zone.

## MEMBRANE FILTRATION EQUIPMENT

Typical membrane equipment components include membranes and support frames, permeate and backpulse pumps, mixed liquor recirculation pumps, programmable logic controllers (PLCs) and controls, air scour blower systems, and a membrane-cleaning system.

### Membrane Configurations

The two main subgroups of MBR configurations are low-pressure immersed membrane systems and high-pressure inline membrane systems. Although inline systems operate at higher flux rates and require less a smaller footprint than immersed MBR systems, immersed systems are more commonly used in municipal applications, as they have significantly lower operating costs and require less frequent cleaning. The focus of this chapter is on low-pressure immersed membrane systems.

In general terms, some types of immersed membranes for use in membrane bioreactors include

- Unsupported hollow-fiber membranes,
- Reinforced hollow-fiber membranes,
- Stationary flat-plate membranes, and
- Rotating flat-plate membranes.

Some additional information on these general types of membranes is included in Chapter 2. However, depending on the manufacturer, the membranes differ in pore size, membrane composition, cassette configuration, operating procedures, and maintenance requirements. Details on specific membrane products must be obtained directly from the manufacturer. (Note: In this publication, the term cassette refers to the largest membrane assembly removable by a crane. Depending on the manufacturer, this may be referred to as a *rack*, a *module*, or a *cassette*.)

Permeation System

The driving force for membrane filtration can be accomplished by either a pumping system or by gravity. External membrane systems can only use a pumping system because the mixed liquor must be pumped through the membranes to generate the transmembrane pressure for filtration (see Figure 3.12). With immersed membranes, permeation is achieved by applying slight suction to draw the clean water from the outside in through the membrane. Gravity siphon systems can be used if the site conditions are suitable, whereas pumped permeation systems can be used in all cases.

There are many possible configurations for permeate pumping systems and many different types of permeate pumps. The simplest configuration is a dedicated permeate pump per membrane train. The permeate pump can be end-suction, centrifugal or positive-displacement, rotary-lobe. Each membrane train is equipped with a permeate header that connects membrane cassettes/racks within the train. When end-suction, centrifugal pumps are used, some means of removing entrained air

**FIGURE 3.12—PERMEATE PUMPS WITH AIR SEPARATORS.**

from the permeate needs to be included to prevent the permeate pumps from losing prime. An air separator connected to a vacuum pump or a Venturi system can be used to remove entrained air.

Because rotary-lobe pumps can handle a higher percentage of entrained air, the permeate header is typically connected directly to the suction side of a self-priming rotary-lobe pump without any air separator in between.

A feature of rotary-lobe pumps is that they can reverse the direction of flow by reversing the direction of rotation of the lobes. Therefore, when rotary-lobe pumps are used in MBR systems, they most often serve double duty as both permeate and the backpulse pumps.

The permeate pump is typically equipped with a variable-frequency drive (VFD). A dedicated magnetic flow (mag) meter and turbidimeter are typically located downstream of each permeate pump.

All permeate pumps typically discharge to a common permeate collection header for downstream disinfection (optional) and discharge.

Backpulse Pumps

Backpulse systems are only included in MBRs systems that use hollow-fiber membranes. These systems are either equipped with a separate backpulse pump or the permeate pump is used for backpulsing (Figure 3.13). During backpulsing, the direction of flow is reversed and the membranes are flushed from the inside out using permeate stored in a backpulse tank. In some systems, the backpulse pump is also used for CIP operations. Separate backpulse pumps are typically end-suction, centrifugal pumps. Centrifugal permeate pumps can also be used for backpulsing without the need for a separate backpulse pump by changing the direction of flow through the use of automatic valves and piping. Reversible self-priming pumps can be used for dual duty—permeation and backpulsing.

Backpulse Tanks

For systems that require backpulsing, a portion of the effluent water is often diverted to a holding tank that is used as the reservoir of backpulse water for both regular and chemically enhanced backpulsing.

FIGURE 3.13—BACKPULSE PUMPS.

Larger MBRs with multiple trains may not require a backpulse tank, as the permeate header is a large enough reservoir of backpulse water. Because only one train at a time is typically in backpulse mode, the supply of permeate water will always exceed the demand if enough trains are in operation.

Cleaning System

Chemical cleaning can range from a fully manual procedure, to a semi-automated system, to a fully automated system. The fundamental task is the same: any membrane train can undergo chemical cleaning without affecting the operation of the other trains (other than increasing flow to the other trains to maintain production). This is achieved by adding (manually or automatically) a chemical solution to the inside and outside of the membrane with the membrane immersed in mixed liquor or cleaning solution or hanging in an empty tank. Cleaning chemicals used and frequency and duration of cleanings are different for the various membrane systems.

Chemical membrane cleanings can be classified into one of two types: maintenance cleaning or recovery cleaning. Maintenance cleaning occurs from once per day to once per week and is fewer than 2 hours in duration. The purpose of maintenance cleaning is to increase time between recovery cleanings. Maintenance cleaning use lower concentrations of chemical than recovery cleans. Recovery cleaning frequency varies among membrane manufacturers and installations but typically ranges from once every 2 to 6 months. The duration of recovery cleaning is between 6 and 24 hours.

A fully automated cleaning system consists of a CIP/backpulse tank, the CIP/backpulse pumps (as described above), a mag meter, and chemical metering stations. Stored permeate from the CIP/backpulse tank is backpulsed through the membrane at a specified flow rate, and the appropriate chemical is injected directly to the permeate header using chemical metering pumps to achieve the desired chemical concentration. In some cases, separate CIP and backpulse tanks are used, and the CIP tank is equipped with a heating system to perform hot water cleaning.

There is a separate metering station for each cleaning chemical used. Cleaning chemicals commonly used include sodium hypochlorite and citric acid. Each metering station is equipped with an appropriate chemical holding tank, a pair of chemical dosing pumps (one on duty, one on standby), and a calibration column. The chemical metering stations are designed to deliver chemical for the following functions:

- Sodium hypochlorite for maintenance cleaning, recovery cleaning, and to flush the CIP/backpulse tank to prevent contamination and biogrowth in the tank and
- Citric acid for maintenance and recovery cleaning.

More information on cleaning procedures can be found in the Maintenance section later in this chapter.

Membrane Air Scour Blowers

The membrane air scour blower system (Figure 3.14) is typically designed as a common group of blowers with installed standby unit(s). Positive-displacement or centrifugal blowers are used. All of the blowers discharge to a common membrane air manifold that delivers air to the air header above each membrane tank (train). Each membrane

FIGURE 3.14—AIR SCOUR SYSTEM.

cassette/rack is connected to the air header above each membrane tank using flexible hose or rigid piping.

The design airflow rate of the air scour blowers is dictated by the membrane manufacturers. Once the air flow rate per membrane cassette/rack is specified, the air scour blower is sized based on the maximum cassette/rack spaces in the tank and the maximum possible liquid level in the membrane tank. To supply the proper airflow rate for the number of membrane cassettes/racks initially installed, blower rates are reduced by VFDs, inlet control vanes, or resheaving the blowers. This approach oversizes the blowers, but it provides the flexibility to add membranes (if required) without having to add blower capacity and it allows sufficient air to the membranes to keep them in production in the unlikely event that the liquid level in the membrane tanks exceeds the normal operating level.

Air may be supplied to the membranes continuously or intermittently (sometimes referred to as "cyclic aeration"). When cyclic aeration is used, the blowers are operated continuously at a fixed speed, and airflow to independent aeration headers is cycled using pneumatically actuated valves. This enables cycling the air between trains or between groups of cassettes within a train. Cycle times can be optimized to minimize energy consumption associated with membrane aeration.

The air scour blowers may form a part of the jet aeration system, in which a two-phase jet system is located at the bottom of each membrane module, introducing both air and mixed liquor to the bottom of the module. The air bubbles blend with the mixed liquor and rise up through the membrane bundle, providing scouring energy to the membrane surface and fluidizing it to prevent solids accumulation.

### Drain Pumps

Some membrane systems require the membrane tank to be drained before a cleaning cycle. The drain pump is typically designed to drain any membrane tank in 30 minutes or fewer. The time to drain a membrane tank is critical, as it is important to minimize the time the membranes are exposed to air to prevent them from drying out.

Typically, a pair of end-suction, centrifugal pumps is used for this purpose. Most systems are designed such that any membrane tank can be isolated and drained without interrupting the operation of the other membrane trains. The drain line from each of the membrane tanks is connected to a common header via an automatic valve. The common drain header is connected to the suction side of the drain pumps. The system is typically designed with the ability to transfer the contents of the membrane tank to either the bioreactor or to sludge wasting.

If the mixed liquor recirculation pumps are dry pit pumps and the system is designed with a dedicated pump per train, these pumps may also function as the drain pumps, eliminating the need for a separate set of drain pumps.

Air Compressors for Pneumatic System Components

These systems are typically designed with a common group of air compressors (one or more on duty, one on standby), each working with a dedicated receiver tank. The system typically includes a common air dryer downstream of the compressors. The compressor and air-dryer system are designed to deliver 550 kPa (80 psig) of instrument air (after pressure-regulating valve) to all pneumatic valve actuators and chemical metering air diaphragm pumps.

Staging Tank

Membrane bioreactor facilities often include a staging tank to provide additional flexibility for membrane handling and maintenance. The staging tank is sized to hold one cassette/rack and is primarily intended for special operations such as membrane cleaning optimization studies and membrane inspections/repair.

# INSTRUMENTATION AND CONTROLS

Programmable Logic Controllers

Each membrane equipment manufacturer assembles its membrane system to include a variety of monitored and controlled instrumentation and equipment. The PLCs in membrane systems provide a variety of critical functions: the monitoring of equipment alarms and setpoints, the trending of operating information such as transmembrane pressure and flow, the fail-safe control and shutdown of equipment, the automated control of certain operating procedures, and the execution of operator-initiated or event-triggered activities.

Membrane PLCs will typically be connected to the plant PLC or supervisory control and data acquisition (SCADA) system for the exchange of operating information and transfer of commands such as the control and sequencing of events that involve both membrane and general plant equipment. Examples of the latter include the coordination of valves, pumps, and gates for the isolation of a membrane tank for a chemical-cleaning procedure.

It is recommended that operators be very familiar with the operation and actions of the PLC system during unusual events such as power failures, maintenance of one or more electrical control panels or PLCs, and high flow events. Returning tanks to service during peak flow events may be automated, but it is still critical that this sequence is understood by operators and carefully monitored for any irregularities.

Turbidimeters

The membrane filtration system is typically designed with a dedicated turbidimeter for each membrane train (Figure 3.15). The turbidimeter is used to monitor membrane integrity by measuring the turbidity in the permeate from each membrane train separately. A small-diameter sampling line from each permeate header carries a continuous flow of permeate from each train to the turbidimeter. The control system is set up with alarms to warn operators if a membrane train exceeds a certain turbidity setpoint. Other setpoints can be set up to automatically shut down and isolate a train if turbidities increased above certain values.

Dissolved Oxygen Meters

Dissolved oxygen meters are often installed in the anoxic and aerobic zones of the bioreactor of a MBR. For smaller systems, where denitrification is not required, dissolved oxygen monitoring may be achieved through handheld devices rather than permanently installed online probes.

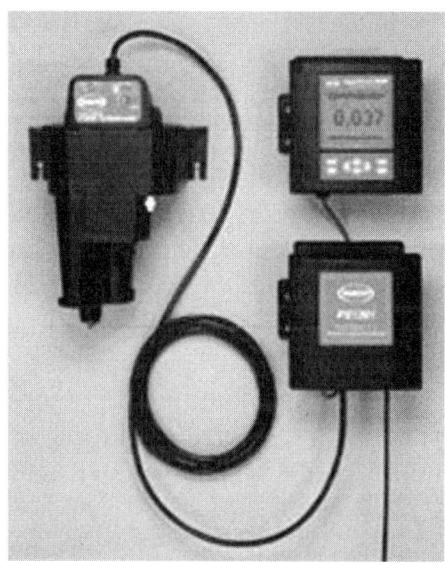

FIGURE 3.15—TURBIDIMETER (COURTESY OF HACH COMPANY).

# Operation

## PRETREATMENT

Because of the numerous pretreatment options available, it is beyond the scope of this publication to anticipate or discuss the operation of various pretreatment facilities such as fine screens or flow equalization tanks (if they exist at the plant). It can only be stated that the proper operation of pretreatment equipment is important for the successful operation and performance of the MBR and its equipment.

## BIOLOGICAL PROCESS OPERATION

The biological activated sludge system is operated much like conventional activated sludge systems, with primary emphasis on controlling the SRT (therefore, the MLSS) and maintaining proper concentrations of dissolved oxygen and nitrate in the various anaerobic, anoxic, and aerobic zones.

Sludge Wasting

In a MBR, the amount of sludge to be wasted can be determined based on maintaining the MLSS concentration within a specified range or keeping the SRT within a specified range.

Sludge wasting can be automated or manually initiated. Either way, the frequency of the wasting can be based on a predetermined period between wasting (e.g., waste every two weeks) or upon obtaining a high MLSS concentration measurement in the membrane compartments. Online monitoring of MLSS is typically not included in MBR plants, but such instruments do exist and have been tested in pilot studies.

Smaller plants may be configured to waste continuously. Once the plant has reached a relatively steady state, regular monitoring of the MLSS concentration in the membrane tanks will indicate whether the wasting rate needs to be increased or decreased.

Sludge can be wasted from a number of locations. The wasting location is determined during the design stage and is typically out of the operator's control. Typical locations for sludge wasting are

- Off the recirculation line, whether this is flowing from the bioreactor to the membranes or vice versa. The volume of sludge wasted will be reduced if the waste sludge is pulled from the line flowing from the membranes to the bioreactor because this line will always have a higher MLSS concentration and, therefore, the mass of solids wasted per unit volume will be higher.
- Off the drain line from the membrane tanks or from the bioreactor. Again, wasting from the bioreactor will result in a smaller overall volume of sludge wasted.
- From the surface of the bioreactor. Surface sludge wasting is the most effective means of controlling foam in a MBR.

When measuring MLSS concentration for the purpose of determining whether it is time to waste activated sludge, the sample location is critical. It is important that samples be taken from both the main bioreactor and membrane compartments. When a membrane manufacturer specifies maximum MLSS concentrations, it is the MLSS concentration in the membrane compartment that is being referenced. With a mixed liquor recycle rate of $4Q$, the MLSS concentration in the bioreactor should be 20% lower than the MLSS concentration in the membrane compartments. If the MLSS concentration in the membrane compartments is high but the MLSS concentration in the bioreactor is low, this is not an indication that sludge needs to be wasted but rather that the mixed liquor recirculation pump rate is too low or there is a problem with one of the pumps.

Dissolved Oxygen

The dissolved oxygen concentration in the aerobic zones should be maintained at 2 mg/L or higher to ensure that oxygen requirements of the microorganisms are satisfied and that nitrification is optimized. The dissolved oxygen concentration in membrane tanks cannot be directly controlled because the airflow to that tank is based on the air scour requirements of the membranes. As a result, the dissolved oxygen in the membrane tanks may be well above 2 mg/L (2 ppm), on the order of 4 to 6 mg/L (4 to 6 ppm). Methods of compensating for the high dissolved oxygen concentrations in the return activated sludge are discussed in the Process Configurations (Biological Treatment) section.

Alkalinity and pH

As in any conventional activated sludge process, the bioreactor pH conditions of a MBR will have a significant effect on the rate of nitrification, with the nitrification rate dropping significantly when the pH is below the neutral range. For stable performance, the pH should be maintained between 6.5 and 8.0 (U.S. EPA, 1993). Because the majority of MBR systems are fully nitrifying systems, alkalinity will be consumed. Alkalinity is consumed at a theoretical rate of 7.1 mg/L (7.1 ppm) of alkalinity (as $CaCO_3$) per milligram per liter (parts per million) ammonium-nitrogen oxidized. Some of this alkalinity will be recovered if the plant denitrifies at a theoretical rate of 3.6 mg/L (3.6 ppm) alkalinity (as $CaCO_3$) per milligram per liter (parts per million) nitrate-nitrogen reduced. However, the net effect will still be a reduction in alkalinity. Coagulant addition for chemical phosphorus removal will typically result in further alkalinity consumption. As a result, many

MBR facilities include a caustic dosing system to maintain the bioreactor pH above 6.5 and ensure that the effluent alkalinity remains above 50 mg/L (50 ppm) as $CaCO_3$.

## MEMBRANE FILTRATION—MODES OF OPERATION

Although the timing of cycles varies from manufacturer to manufacturer, in general, the membrane filtration of a MBR cycles between permeating and relaxing or permeating and backpulsing (Figure 3.16). Some manufacturers recommend periodic maintenance cleaning and, eventually, every membrane system will need to undergo a full chemical recovery clean.

Permeation

Permeate flow control is typically based on the plant influent flowrate (as measured by the influent flow meter), the water level in the membrane or bioreactor tank, and/or the level in the flow equalization tank (if used at the facility). The permeate flow is controlled by the PLC and the VFD on the permeate pumps (or opening and closing control valves for gravity/siphon systems without permeate pumps). Often, the maximum permeate flow setpoint is limited by the transmembrane pressure (TMP) so that the membranes are not exposed to excessive TMP that might cause increased fouling.

Relax

A membrane train is said to be in relax mode when filtration is suspended (no permeate flow) and the air scour is left on. Typically, a membrane train is relaxed for 30 seconds to 1 minute out of every 10 to 15 minutes or more, depending on the manufacturer. By aerating the membranes without permeating, some of the solids that have accumulated on the membrane surface are dislodged. The result is a reduction in TMP when the next permeate cycle begins. To maintain the design flux, an instantaneous flux must be calculated and converted to a flow. For example, to relax membranes for 1 minute out of every 10 and maintain design (net) flow (or flux), the gross (or instantaneous) flux must be 10% higher than the design flux. So, if the design per-

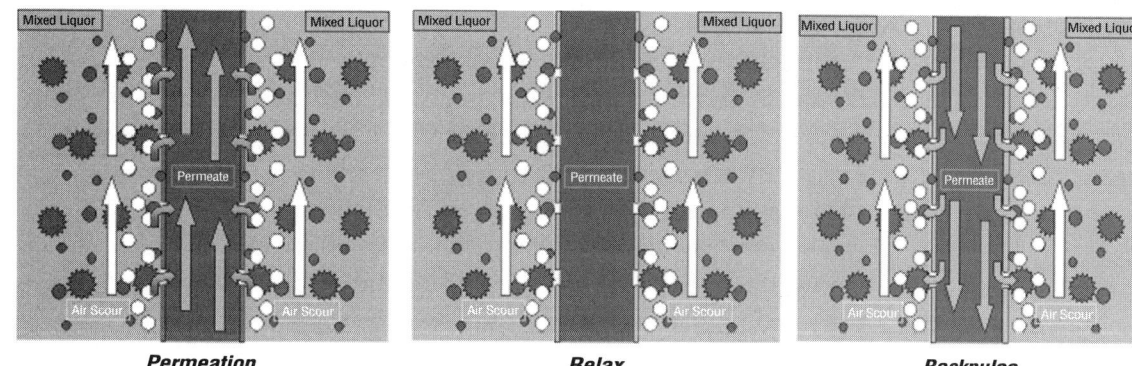

**FIGURE 3.16—MODES OF MBR OPERATION.**

meate flow rate is 38 L/min (10 gpm), in practice it will be 42 L/min (11 gpm). At some point, relaxing the membrane assembly will not recover the design flow at a reasonable TMP and a backpulse, chemical maintenance clean, or recovery clean must be performed.

Backpulse/Backwash

Backpulsing refers to reversing the flow of permeate through the membranes to flush trapped particles from the membrane pores and cavities. Hollow-fiber membranes can be backpulsed while flat-sheet membranes cannot because of the way they are made and attached to the membrane plates. For hollow-fiber systems, the fibers tend to enlarge during backpulsing (and collapse marginally during permeation) so that the cavities are larger and debris is readily dislodged.

Chemically Enhanced Backpulse/Backwash (Maintenance Clean)

Maintenance cleaning (sometimes referred to as *chemically enhanced backpulsing*) typically refers to the addition of chemicals during a backpulse cycle to provide additional cleaning and increase flux rates more frequently. Hypochlorite can be used to control organic growth, whereas citric acid is useful if long-term deposition of inert materials is a concern. More detail on maintenance cleaning is provided in the Maintenance section.

### Recovery Cleaning

Although the rate of fouling can be controlled using the techniques described above (i.e., air scouring, relax, backpulse, and maintenance cleaning), it is normal for membranes to become fouled over the course of operation to such an extent that permeability can only be recovered through the use of an extended chemical cleaning. Recovery cleaning can be an event-triggered operation—initiated when performance drops below a certain level (based on TMP and/or permeability)—or time-triggered operations—initiated every 6 months regardless of whether performance has decreased significantly. Most manufacturers recommend performing maintenance cleaning based on time rather than waiting for performance to decrease. The reason for this is twofold. First, the rate of fouling has a tendency to increase rapidly once a certain permeability threshold is reached. The rate of membrane fouling is highest during high flow conditions, and taking additional trains offline for cleaning during a high flow period when membrane performance is already on the decline will only exacerbate the situation. Secondly, particularly during the first few years of operation, membrane fouling is more difficult to detect when fluxes are low, as only minute changes in TMP are observed. Under these conditions, operators may be lulled into a sense of security. However, the true condition of the membranes becomes evident when the wet-weather season approaches and the system is required to operate at peak capacity. More detail on recovery cleaning procedures can be found in the Maintenance section.

### Flow Control

There are two primary control philosophies behind the flow control of a MBR. Plant output will either be based on the influent flow or a desired production setpoint; the latter is primarily true for "scalping plants" whose primary objective is the production of reuse water rather than wastewater treatment.

When plant output is based on influent flow, permeate flow will be controlled either to maintain a certain liquid level in the bioreactor (or membrane tank or equalization tank) or match the influent flow rate.

When all trains are operating at their lowest flows and a low-level setpoint is reached, individual trains begin to go into standby sequentially until just enough trains are operating to maintain the desired capacity. It is not uncommon for the entire plant to be in standby mode during the lowest flow periods of the day.

When trains are operating at their maximum capacity and a high-level setpoint is reached, the system is programmed to verify if any trains are operating in a mode other than "permeation", and returns them to "permeation" mode if possible. Any maintenance or recovery cleaning procedures are aborted, and any trains in standby are put into service.

## MEMBRANE AERATION

Air scouring is used by immersed membrane systems to cause shear forces and turbulence across the surface of the membranes to keep solids away from them and maintain optimum conditions for flow. Most air scour systems operate continuously, although some systems cycle the air between groups of membranes. If the air scour system fails, then the TMP will rise quickly, possibly causing alarm conditions and the need for cleaning to restore normal operation.

## ALARMS AND TRENDS

Transmembrane pressure alarms and trends are important to warn operators that membranes are fouling and a maintenance or recovery cleaning procedure may soon be required and should be planned. The TMP changes with each backpulse or relax cycle; therefore, the operator can use long-term trending information to assess the rate of fouling of the membranes.

Other critical alarms and trends that should be closely monitored include

- High and low tank levels,
- High flows,
- High effluent turbidity,
- Low membrane scour airflow, and
- Low mixed liquor recycle flow.

## OPERATIONAL CHANGES AS A RESULT OF EVENTS AND SEASONAL CHANGES

Seasonal changes in operation can be expected at plants that receive high flows during spring snowmelt conditions or periods of heavy rain. Also, loading to the plant may peak as a result of seasonal food processing discharges (e.g., processing of apple crops in the fall in the United States) or other seasonal events. Operators should carefully consider seasonal and other loading events in their scheduling of membrane maintenance, recovery cleaning, and tank inspection activities.

## CAPACITY MANAGEMENT

Equipment redundancy is typically provided as part of the system, which means that the operator has the ability to take some equipment out of service for maintenance during most times of the year. Operators should be familiar with the level of redundancy provided at the plant, and determine whether it is possible to have a tank or equipment train out of service during the peak day, peak week, or peak month flow or load condition. Care should then be taken to ensure that any operator-initiated activities that would remove equipment from service are performed when anticipated flows are within the capacity of the remaining equipment.

In addition to standby equipment, backup PLCs and power generators are often included. At a minimum, backup power is required for fine screens, membrane air, permeate pumps (if included), back-pulse pumps, and controls to allow incoming flows to be carried through the plant. Because most membrane plants are pumped systems, gravity flow during power outages is not typically possible.

## OPTIMIZATION

Optimization of some plant processes may be possible at MBR facilities, in particular to minimize the consumption of energy and cleaning chemicals.

Air scour blowers are a significant source of operating energy consumption, with energy requirements sometimes approaching that of process aeration blowers. Air scour blowers must operate to support membranes that are in service; therefore, the operator should be alert to trends in plant flow that may allow portions of the membrane equipment to be removed from service for prolonged periods of time.

Permeate pumps are another significant user of energy, with the running cost related to both permeate flow and TMP. The TMP required is primarily related to the degree to which membranes are fouled. However, it is also a function of the operating MLSS concentration, system SRT, and effectiveness of automated and scheduled membrane cleaning operations. These latter conditions are within the control of the operator and could be adjusted to optimize (minimize) energy consumption by permeate pumps. Lowering the MLSS concentration in a hollow-fiber system, for example, as long as the SRT is still within the range required for effective treatment, will result in reduced TMP and reduced energy consumption.

Chemical use is another source of operating cost. The quantity of chemical required is a function of the effectiveness of the cleaning system and air scour and backpulse/relax operations. The most effective combination of chemical use for maintenance and recovery cleaning can be determined through operator experience.

Chemical use for phosphorus removal should also be minimized. Although effluent quality will not be negatively affected by overdosing coagulants, the cost of chemicals can become excessive if dosing rates are not monitored with effluent quality.

# Routine Monitoring

## PRETREATMENT

The operator should ensure that pretreatment systems such as flow equalization tanks (if provided), fine screens, and grit-removal systems are functioning as intended.

## BIOLOGICAL TREATMENT

Proper operation of the activated sludge process is required to ensure compliance of effluent quality with permit requirements for soluble constituents such as biochemical oxygen demand, ammonia, nitrate, total nitrogen, and orthophosphate. Proper process operation is typically also required to ensure proper operation of membrane equipment in terms of capacity, flux, and TMP. A testing regimen similar to that shown in Table 3.2 is recommended.

The plant SRT should be calculated daily, as this is the primary indicator of process operation and stability. For plants that include anoxic zones for denitrification, the dissolved oxygen in the mixed liquor recycle and the dissolved oxygen and nitrate concentrations

Table 3.2  Typical MBR System Testing Requirements.

| Parameter[a] | Minimum Test Frequency[b] | Recommended Test Frequency | Sample Location[c] IN/EFF/MBR[d] | Notes |
|---|---|---|---|---|
| TSS/VSS | 1/week | 1/day | IN/MBR | Method 2540 D[e] |
|  |  |  |  | Method 2540 E |
| Temperature | 1/day | Continuous | MBR |  |
| pH | 1/day | Continuous | IN/EFF/MBR | Method 4500-H B |
| Dissolved oxygen | 1/day | Continuous | MBR (multiple locations) | Method 4500-O G |
| Viscosity | 1/week | 1/day | MBR | See manufacturer's protocol |
| Filterability | 1/week | 1/day | MBR | See manufacturer's protocol |
| Turbidity | 1/week | Continuous | EFF | Method 2130 B |
| $BOD_5$ | 2/month | 1/week | IN/EFF | Method 5210 B |
| COD | 1/week | 2/week | IN/EFF | Method 5220 |
| TKN | 2/month | 2/week | IN/EFF |  |
| $NH_3$-N | 2/month | 2/week | IN/EFF | Method 4500-$NH_3$ |
| $NO_3$-N | 2/month | 2/week | EFF | Method 4500-$NO_3$ |
| Orthophosphate | 2/month | 2/week | IN/EFF |  |

[a] TSS = total suspended solids; $BOD_5$ = five-day biochemical oxygen demand; COD = chemical oxygen demand; TKN = total Kjeldahl nitrogen; $NH_3$-N = ammonia-nitrogen; and $NO_3$-N = nitrate-nitrogen.

[b] Parameter type and testing frequency assume typical municipal waste. For industrial or other applications, requirements can change.

[c] IN/EFF/MBR refer, respectively, to influent wastewater downstream of pretreatment, permeate downstream of membranes only (not postdisinfection), and mixed liquor inside the MBR.

[d] Collect 24-hour composite samples for influent and effluent testing. For mixed liquor laboratory samples, simply submerge a container directly into the MBR. Follow sampling protocol as required for laboratory analysis of dissolved oxygen.

[e] Listed methods are per *Standard Methods for the Examination of Water and Wastewater* (latest approved edition).

within the anoxic zones should be measured routinely, as these are primary indicators of the performance and operation of the anoxic zone. If improved denitrification is needed, the operator should attempt to prevent the transfer of excess dissolved oxygen to the anoxic zone either by reducing the dissolved oxygen in the recycle or by reducing the recycle flow rate. Note that the recycle flow rate from the membrane tanks and the dissolved oxygen in that recycle flow are limited by membrane equipment operating requirements for air scour and recirculation flow; therefore, the manufacturer's recommendations must be followed. For plants with low total nitrogen requirements, other means of controlling dissolved oxygen in the recycle to the anoxic zone may be considered, such as including either a separate recycle line to allow independent optimization and control of solids and nitrogen or a separate deoxygenation zone to reduce dissolved oxygen in the recycle stream before it enters the anoxic zone. If the plant also performs biological phosphorus removal, measurement of orthophosphate concentrations in the anaerobic, anoxic, and aerobic zones is recommended as well as measurement of nitrate in the influent to the anaerobic zone. If improved phosphorus removal is desired, then the operator should attempt to reduce the transfer of nitrate (and dissolved oxygen) to the anaerobic zone either by reducing the recycle flows or ensuring that the anoxic zones are removing as much of the nitrate as possible. The operating and maintenance manual for a specific plant should provide additional options for optimizing the treatment process.

In addition to monitoring effluent quality, mixed liquor filterability should be monitored at least twice per year. Time-to-filter, capillary suction time (CST), and viscosity test procedures should be followed per the latest U. S. Environmental Protection Agency (U. S. EPA)-approved version of *Standard Methods for the Examination of Water and Wastewater*, which is copublished by American Public Health Association, American Water Works Association, and Water Environment Federation.

# MEMBRANE SYSTEMS
### System Performance
Instantaneous and historical readings of permeate flows are typically available either directly from the control panel's human machine

interface or on a personal computer based SCADA system. Daily average, peak hourly, and peak instantaneous flows, with all flows measured as net flows corrected for backpulse and cleaning requirements, should be recorded.

Long-term permeate flows, including moving averages of weekly and monthly flows, are important for monitoring membrane performance and recovery from periods of sustained high flux and flow requirements.

Mixed liquor suspended solids concentration in the membrane tank should be measured daily because membrane flux is reduced by high MLSS.

Transmembrane pressure should be tracked continuously and trended. Reductions in TMP should be observed after each relax mode or backpulse cycle. Over time, the TMP may gradually increase, indicating that a maintenance or recovery cleaning is required.

Turbidity should be continuously monitored and trended. The permeate turbidity should not change and is typically unaffected by TMP or flux within design ranges. A sudden increase in turbidity, however, can indicate breakage or damage to a membrane sheet or fiber or a permeate pipe fitting. Any increase should, therefore, be a signal to the operator to perform an integrity test to identify any broken or damaged membranes or piping connections.

If the system is operated in a cycle that includes regular backpulsing, routine permeability monitoring during membrane backpulsing can be used as an indicator of membrane surface fouling. A significant decrease in backpulse permeability compared with that of membranes when initially put into service is important to note because it may be an indication of

- Surface fouling of the membranes;
- The presence of particulate matter in the backpulse water; or
- A backpulse-piping integrity issue, resulting in mixed liquor entering the backpulse lines.

### Equipment and Instrumentation

Membrane manufacturers will provide detailed instructions related to monitoring equipment and data provided by the instrumentation system.

Most of these monitoring activities can be automated so that alarms are generated to notify operators when a parameter exceeds a certain value. The plant monitoring plan recognizes that the duty of the operator is to operate the complete plant, of which the membrane system is only one component. If operators ensure that all other treatment-system components are operating properly, this will improve performance and minimize the maintenance required on the membrane system. Trended data and alarm notification are two important information sources provided by membrane manufacturers to assist the operator.

# Maintenance

## *CLEANING*

Biological growth coats the outer layer of membranes in a submerged MBR during operation of the system. Finer particles may penetrate pores of the membrane, causing an increase in pressure loss. Both biomass coating and particle penetration are known as organic fouling because the nature of the material causing the low restriction is organic in nature and can be controlled by chlorination. Additionally, some chemical precipitation can occur on membrane surfaces, also affecting pressure loss and flow capacity of the membrane. This is known as inorganic fouling, or inert fouling, and can be controlled by occasional treatment with an acid to remove precipitated inert material. During the operation of the MBR system, continuous membrane fouling control methods such as air scouring, relaxing, and/or backpulse are used but, periodically, more aggressive cleaning methods are required to control membrane fouling.

Maintenance Cleaning

Maintenance cleaning should be performed as recommended by the manufacturer; cleaning cycles are intended to provide interim cleaning of the membrane system between scheduled recovery chemical cleaning. This extends the time between recovery chemical cleanings. Although they are not as comprehensive as full recovery chemical cleanings and do not involve taking tanks out of service for significant periods of time, they are designed to be more effective than regular backwashes at removing particles from the membrane surface.

Maintenance cleaning can be performed as often as once per membrane tank per day. However, one maintenance cleaning per membrane tank per week is a more common frequency. Typically, chlorinated water is used. If inorganic fouling is suspected, then a weak acid solution (e.g., citric or oxalic acid) may also be used as recommended by the manufacturer.

Sodium hypochlorite (NaOCl) solution concentrations typically used for maintenance cleaning range from 100 to 500 mg/L (100 to 500 ppm).

Citric acid solution concentrations typically used for maintenance cleaning range from 1000 to 5000 mg/L (1000 to 5000 ppm).

Specific maintenance cleaning procedures vary from manufacturer to manufacturer, but general steps typically include

(1) Isolate the train to be cleaned. Permeate and mixed liquor recycle through the train are stopped. Air scour may continue, depending on the manufacturer's recommendation.

(2) Drain the membrane tank. Some manufacturers' maintenance cleaning procedures require that the tank be drained before chemical injection; others are performed in a full tank.

(3) Conduct a chemically enhanced backpulse/backwash cycle. Backpulse chemical solution (chlorinated water or acidic solution) through the membranes at normal backpulse flow (or higher as specified by the manufacturer). The duration of the initial backpulse depends on the piping layout for the specific project because the purpose of the initial backpulse is to flush the permeate piping and fill it with chemical solution.

(4) Relax the system for 5 to 15 minutes.

(5) Conduct another chemically enhanced backpulse/backwash cycle. Backpulse chemical solution (chlorinated water or acidic solution) through the membranes at normal backpulse flow (or higher as specified by the manufacturer) for 30 seconds to 2 minutes.

(6) Repeat steps 4 and 5 as specified by the manufacturer. The total duration of maintenance cleaning can range from 30 to 60 minutes per train.

Recovery Cleaning

Ultimately, a more aggressive cleaning will be required to recover membrane permeability. It is normal for permeability to decrease rapidly during the first few months of operation, and recovery to "as-new" membrane conditions should not be expected following a recovery cleaning. The intention of recovery cleaning is to improve membrane permeability to >80% of the permeability observed following the initial membrane "break-in" period. Contract documents and operating manuals should be consulted to determine specific criteria used to define the success or failure of a recovery cleaning.

In general, a recovery cleaning should be conducted when recommended by the manufacturer, which is typically when the TMP or permeability changes by a prescribed amount or after a defined period of operation such as 6 months. There are several factors that can cause the TMP to increase at a given flux; therefore, the troubleshooting guide later in this chapter should be consulted before a cleaning is performed.

Specifically, a recovery cleaning is used to dislodge particles from the microstructure of the membranes that have accumulated over time. The type of particle to be removed dictates the chemical that should be used during a recovery cleaning cycle. Unless a history of inorganic scaling has been documented, always perform a bleach cleaning first to remove organic fouling. If permeability does not fully recover, perform a second cleaning using a reductive acid solution to remove scaling.

The operator has some discretion as to when chemical cleanings are performed and what chemicals will be used for each cleaning cycle. Some situations that will influence the time to initiate a cleaning include

- *Transmembrane pressure of the modules in a membrane tank:* the maximum desired TMP across membranes in a tank is set by the manufacturer, but it is always better to clean membranes before they reach this point.
- *Transmembrane pressure of the modules in other membrane tanks:* when sufficient capacity exists in other operating basins, the operator has the flexibility to schedule cleaning cycles at a period of low flow. It is good operational practice to start

membrane tanks up one at a time during initial startup to stagger the timing of the chemical cleanings.

- *Influent flow to the plant:* the operator should try to avoid scheduling chemical cleanings at periods of peak flow.
- *Available equalization capacity:* chemical cleanings can only be performed if sufficient equalization is available to store untreated wastewater for the duration of the cleaning. Sufficient equalization basin volume must be available before beginning the cleaning cycle.
- *Water temperature:* chemical cleanings are most effective when the cleaning solution is above 20° C (68° F). Heaters are often included to heat the chemical cleaning solution in regions when winter water temperatures are expected to be below 10° C (50° F) for extended periods of time. If possible, recovery cleanings should be scheduled during periods when the chemical clean water is warmer to reduce or eliminate the need to heat the cleaning water. A combination of higher chemical concentrations and longer contact times can be used to compensate for lower temperatures. The operator will need to consult with the individual membrane manufacturer before increasing either the concentration or duration of a recovery cleaning, as each manufacturer specifies different maximum chemical contact times for various cleaning chemicals.

Specific recovery cleaning procedures vary from manufacturer to manufacturer, but general steps may include

(1) Determine the type of chemical cleaning required using the manufacturer's instructions, with typical guidelines in Table 3.3 below. Prepare the necessary amount of cleaning solution to perform a cleaning. Use either potable water or permeate water for dilution purposes. Note: the cleaning solution may also need to be heated.

(2) Suspend the filtration (permeate operation) of the membrane assemblies to be cleaned. Depending on the type of system, this may require isolation of membrane tanks, turning off pumps, and/or closing permeate valves.

Table 3.3  Recovery Cleaning Chemical Selection Chart.

| Suspected Fouling Material | Recommended Cleaning Chemical | Alternate Cleaning Chemical | Cleaning Solution Concentration | Cleaning Duration |
| --- | --- | --- | --- | --- |
| Organic | Sodium hypochlorite (bleach)[c] | Hydrogen peroxide | 500–5000 mg/L | 6–24 h |
| Aluminum oxide[a] | Oxalic acid[d] | Citric acid | 1000–10 000 mg/L | 6–24 h |
| Ferric oxide[a] | Oxalic acid[d] | Citric acid | 1000–10 000 mg/L | 6–24 h |
| Calcium carbonate[b] | Hydrochloric acid | Citric acid | 1000–10 000 mg/L | 6–24 h |

[a] Oxides may form in the presence of common coagulants such as alum and ferric chloride.
[b] Scaling may occur in the presence of hard wastewater.
[c] The concentration of bleach can affect system biology. Therefore, be sure to adjust the cleaning solution concentration at MLSS concentrations <10 000 mg/L (<10 000 ppm).
[d] To avoid the formation of the mineral scale calcium oxalate, do not use oxalic acid in the presence of hard water.

(3) Continue to aerate for a period of time (15 to 60 minutes) to dislodge accumulated solids from the membranes.

(4) At this point, the tank may be drained, depending on the manufacturer's instructions. For some smaller plants, the cassette/rack is removed from the system, and cleaning continues on individual cassettes in the staging tank rather than in the main membrane compartment. For a CIP cleaning procedure, membranes remain in their working position for the recovery cleaning. These facilities are designed with multiple membrane trains to ensure continuous treatment of wastewater even while one is out of service during the cleaning cycle. In these plants, each membrane tank must be equipped with isolation valves or gates and with a tank drain system. With these provisions, an entire train of the membrane system can undergo recovery cleaning as a unit.

(5) Depending on the manufacturer, backpulse or pour the dilute cleaning solution into each membrane cassette/rack individually

or at a tap on the main permeate header. Chemical addition typically takes between 5 and 10 minutes, with a maximum time of 20 minutes.

(6) Allow the membrane cartridges to soak in the cleaning solution for 6 to 24 hours.

(7) The membranes will now be sitting in a relatively highly concentrated chemical solution. The spent cleaning solution needs to be disposed, and the method of disposal will vary, depending on the manufacturer and overall plant layout. One option is to return the spent cleaning solution back to the bioreactor at a controlled rate to eliminate adverse affects on the biological process.

Physical Cleaning

Pretreatment of wastewater to remove screenable materials and grit has already been emphasized. Even with excellent pretreatment systems, it can be expected that some screenings will still be present within the plant and that some debris will find its way into membrane tanks. Routine draining and cleaning of membrane tanks is, therefore, an important activity that could be conducted on an annual or semiannual basis. Tanks should be cleaned of any debris, the condition of any wall coatings should be inspected, and the integrity of frames and assemblies that support membranes should be verified. In all cases, manufacturers' instructions should be followed with respect to the draining of the tank and possible exposure of membranes to sunlight, air, and heat.

A uniform distribution of cleaning air is required to prevent membrane fouling and maintain a given flux. Therefore, routine maintenance of membrane assembly diffusers is also recommended.

## MEMBRANE INTEGRITY

Integrity Testing

The primary means of determining membrane integrity in a MBR is by turbidity of the permeate. The operating practice of using turbidity readings to identify damaged membranes is described in the next section.

A more direct method of measuring membrane integrity is by using a pressure decay test (PDT). In a PDT, the lumen side of the fibers is pressurized to 34 to 62 kPa (5 to 9 psi), and all valves are shut to keep it pressurized. The PDT pressure must be below the bubble point of the wet membranes so that air will not flow through normal pores in the fiber walls beyond what is a result of diffusion. If any decay in pressure is noted above a specified limit, this indicates a loss of integrity in the fibers (broken fibers or large pores) or a leak in filtrate piping. Determining the source of the integrity loss can be achieved by visually inspecting the water surface in the tank and identifying the location of any bubbling. This visual inspection can be best conducted in a separate tank containing clean water. Note: if the fibers are dry, then air can flow through the fiber walls and will give a falsely high-pressure decay rate.

Online PDT is commonly used in water treatment plants; however, it is not practical in a wastewater treatment plant. The level of detection is far beyond that which is required to maintain effluent quality requirements of any wastewater treatment plant. In-tank testing is not feasible, as the mixed liquor will make determining the source of bubbles almost impossible.

For a MBR using hollow fibers, a small hole in a membrane fiber will likely not have any measurable effect on effluent quality. With thousands of fiber strands, the contribution to turbidity of any single fiber is minute. Additionally, because of the elevated concentration of solids in the process tank and the small diameter of the membrane fibers, even a fiber that is completely severed will quickly plug up with solids.

Note that bubble testing is not applicable to flat-sheet membrane equipment; the operator should follow the manufacturer's recommendations in these cases.

Identifying and Repairing Damaged Membrane Elements

The permeate turbidity should be approximately 0.2 NTU or less for newly installed membrane equipment. If a membrane sheet or fiber becomes damaged or broken, the first indication to the operator will be an increase in permeate turbidity. Some systems use clear plastic tubing connecting individual cassettes/racks to the permeate header. Visual inspection of the tubing is one method of narrowing down the source of the high turbidity. If the system does not include clear tubing or it is difficult to differentiate water quality based on visual inspection alone, then the following general procedures can be used to identify and repair the damage:

(1) Sequentially measure the turbidity of each membrane tank or train to identify the tank within which the damaged membrane(s) is located.

(2) Within the identified tank and with the tank in operation, sequentially isolate each membrane assembly by shutting off its permeate valve and observe the resulting turbidity for the remaining assemblies still in operation in that tank. A decrease in turbidity when an assembly is isolated (valved off) is an indication of damage to the membranes in that assembly.

(3) Depending on the type of membrane and the manufacturer's instructions, isolate or remove the membrane assembly containing the damaged membranes.

(4) If sheet membranes are installed, inspect the membranes for damage and repair according to the manufacturer's instructions.

(5) If fiber membranes are installed, place the assembly into a dip tank or repair tube and apply air to the permeate side of the membrane (according to manufacturer's instructions). Identify the problem fiber by observing bubbling from the broken fiber through the water surface.

(6) Seal the fiber according to the manufacturer's instructions.

(7) Release air pressure.

(8) Reinstall assembly and return to service.

# INSTRUMENT CALIBRATION

Proper operation of membrane equipment and of MBRs depends on the reliable operation of the various instrumentation and automated control systems. Maintenance and calibration of the system instrumentation are, therefore, critical both for the performance of the plant and for the protection of the membrane equipment. Some examples include

- The influent plant flow in combination with liquid levels in membrane tanks, biological reactor tanks, and/or the mixed liquor channels is often used to control the permeate pump flow rate.
- The permeate flow meter is used to automatically calculate the flux on membranes. There may be limits on the flux that can be maintained for 1 hour, 1 day, or 1 month; therefore, tracking of the flux is important to equipment operation.
- The TMP is used to protect membrane equipment by preventing the permeate pump from attempting to draw high flows that would require excessive TMP.
- Mixed liquor temperature, in combination with the flux and TMP, is used to calculate membrane permeability. This is a standardized capacity parameter that is useful for long-term monitoring of membrane performance.
- Turbidity of the permeate is a primary indicator of membrane integrity. An increase in turbidity might result from broken fibers or sheet membranes or from faulty piping connections. In either case, the turbidity sensor is used by the operator as an early indicator of the need for physical inspection and maintenance of the membrane system.
- Other flows and levels are also critical, such as those associated with chemical feed systems.

# Troubleshooting

This section presents a table (Table 3.4) that shows various tests and responses for troubleshooting the operation of the biological process and the membrane equipment when the treatment system is not performing as intended.

Table 3.4  Membrane Bioreactor Troubleshooting.

| | PROBLEM | POSSIBLE CAUSE | REMEDIES |
|---|---|---|---|
| 1 | Air scour rate fails to reach the specified value. | Damaged blower(s). | Repair or replace the blower(s). |
| | | Malfunctioning diffuser(s). | Repair the diffuser(s). |
| 2 | Uneven aeration of an individual membrane assembly. | Diffuser clogging. | Wash or clean the diffuser(s). |
| | Uneven aeration of multiple membrane assemblies. | Membrane unit diffusers are not level. | Level the diffusers. |
| 3 | High filtration pressure or desired operating flux cannot be obtained. | Desired operating flux or flowrate is higher than design. | Adjust operational flux or flowrate to match the design flux or flowrate. |
| | | Membrane surface is fouled. | Aerate for 10 minutes without filtration. If the flux rate does not recover, then carry out maintenance cleaning or recovery cleaning. Note that if the MLSS concentration is high, then the cause may be solids deposition, in which case the tank should be drained and the membranes washed, and the MLSS should be reduced. |
| | | Characteristics of the activated sludge are bad. | Perform a CST test to determine if sludge filterability has changed. |
| | | ■ MLSS too high<br>■ SRT too low (below that needed for nitrification) | Improve the characteristics of the sludge by adjusting the activated sludge process operation. |

Table 3.4  Membrane Bioreactor Troubleshooting (*continued*).

| Problem | Possible Cause | Remedies |
|---|---|---|
| | ■ Insufficient dissolved oxygen in aerobic zones | |
| 4 The permeate is cloudy and/or the permeate turbidity is higher than desired. | The membrane surface becomes hydrophobic. | Treat the membrane with a hydrophilic agent. |
| | The permeate tube or connections have separated or are damaged. | Repair and replace the tube or connections. |
| | The membrane fibers or sheets are damaged. | Repair or replace the membrane fibers or sheets. |
| | Damage to the membrane assembly manifold. | Repair or replace the manifold. |
| | Permeate piping connection is loose. | Repair the permeate piping connection. |

CHAPTER FOUR

# Low-Pressure Membranes for Effluent Filtration

Introduction . . . . . . . . . . . . . . . . . . . . . . . . . . . . . . . . . . . . . . . . . . . . . . . 115
    Applications . . . . . . . . . . . . . . . . . . . . . . . . . . . . . . . . . . . . . . . . . . . 115
    Feed Water Quality . . . . . . . . . . . . . . . . . . . . . . . . . . . . . . . . . . . . . 116
    Filtrate Water Quality . . . . . . . . . . . . . . . . . . . . . . . . . . . . . . . . . . . 117
    Treatment Mechanism . . . . . . . . . . . . . . . . . . . . . . . . . . . . . . . . . . 118
    Safety . . . . . . . . . . . . . . . . . . . . . . . . . . . . . . . . . . . . . . . . . . . . . . . . 120
Process Configurations . . . . . . . . . . . . . . . . . . . . . . . . . . . . . . . . . . . . . 120
    Pretreatment . . . . . . . . . . . . . . . . . . . . . . . . . . . . . . . . . . . . . . . . . . 121
    Membrane System . . . . . . . . . . . . . . . . . . . . . . . . . . . . . . . . . . . . . 123
    Chemical Storage and Feed Systems . . . . . . . . . . . . . . . . . . . . . . . 124
    Posttreatment . . . . . . . . . . . . . . . . . . . . . . . . . . . . . . . . . . . . . . . . . 124
    Residuals Handling . . . . . . . . . . . . . . . . . . . . . . . . . . . . . . . . . . . . . 125
Equipment Configurations . . . . . . . . . . . . . . . . . . . . . . . . . . . . . . . . . . 126
    Pretreatment Components . . . . . . . . . . . . . . . . . . . . . . . . . . . . . . . 126
    Membrane Equipment Components . . . . . . . . . . . . . . . . . . . . . . . 128
Operation . . . . . . . . . . . . . . . . . . . . . . . . . . . . . . . . . . . . . . . . . . . . . . . . 128
    Pretreatment . . . . . . . . . . . . . . . . . . . . . . . . . . . . . . . . . . . . . . . . . .129
    Automated Systems . . . . . . . . . . . . . . . . . . . . . . . . . . . . . . . . . . . . 131
        Filtrate Flow Control . . . . . . . . . . . . . . . . . . . . . . . . . . . . . . . . 132
        Transmembrane Pressure Alarms . . . . . . . . . . . . . . . . . . . . . . . 133
        Backwash System . . . . . . . . . . . . . . . . . . . . . . . . . . . . . . . . . . . 133

|     Productivity Management . . . . . . . . . . . . . . . . . . . . . . . . . . . . . . . . . . . . . . . . . . 133
        Pretreatment . . . . . . . . . . . . . . . . . . . . . . . . . . . . . . . . . . . . . . . . . . . . . . 135
        Design Operating Flux . . . . . . . . . . . . . . . . . . . . . . . . . . . . . . . . . . . . . . 135
        Backwashing Frequency . . . . . . . . . . . . . . . . . . . . . . . . . . . . . . . . . . . . 136
        Clean-In-Place Sequence . . . . . . . . . . . . . . . . . . . . . . . . . . . . . . . . . . . 136
        Seasonal Changes in Operation . . . . . . . . . . . . . . . . . . . . . . . . . . . . . . 137
Routine Monitoring . . . . . . . . . . . . . . . . . . . . . . . . . . . . . . . . . . . . . . . . . . . . . . 137
    Pretreatment System . . . . . . . . . . . . . . . . . . . . . . . . . . . . . . . . . . . . . . . . . . 137
    Membrane System . . . . . . . . . . . . . . . . . . . . . . . . . . . . . . . . . . . . . . . . . . . 138
    Test Parameters . . . . . . . . . . . . . . . . . . . . . . . . . . . . . . . . . . . . . . . . . . . . . 138
    Data Analysis and Reporting . . . . . . . . . . . . . . . . . . . . . . . . . . . . . . . . . . 139
Maintenance . . . . . . . . . . . . . . . . . . . . . . . . . . . . . . . . . . . . . . . . . . . . . . . . . . . 140
    Chemical Cleanings . . . . . . . . . . . . . . . . . . . . . . . . . . . . . . . . . . . . . . . . . . 141
    Integrity Testing . . . . . . . . . . . . . . . . . . . . . . . . . . . . . . . . . . . . . . . . . . . . 141
    Tank Washing and Cleaning . . . . . . . . . . . . . . . . . . . . . . . . . . . . . . . . . . . 142
    Instrument Calibration . . . . . . . . . . . . . . . . . . . . . . . . . . . . . . . . . . . . . . . 144
Troubleshooting . . . . . . . . . . . . . . . . . . . . . . . . . . . . . . . . . . . . . . . . . . . . . . . . 145

# Introduction

This chapter presents information on the use of microfiltration (MF) and ultrafiltration (UF) membranes as tertiary filters at a municipal or industrial wastewater treatment facility, including their use to provide pretreatment for reverse-osmosis (RO) membranes.

## *APPLICATIONS*

In municipal wastewater treatment plants, low-pressure membranes are used to provide tertiary treatment for secondary effluent to improve the quality of the water for subsequent use. Filtrate from low-pressure membranes is suitable for many reuse applications, including public access spray irrigation, discharge to surface waters, pretreatment for RO feed water, and the provision of high-quality water for industrial applications. A 2003 survey of full-scale wastewater treatment facilities using membranes found the largest use of low-pressure membranes for tertiary filtration is to pretreat secondary effluent to facilitate further treatment by RO before aquifer recharge or other forms of indirect potable reuse (see Figure 4.1).

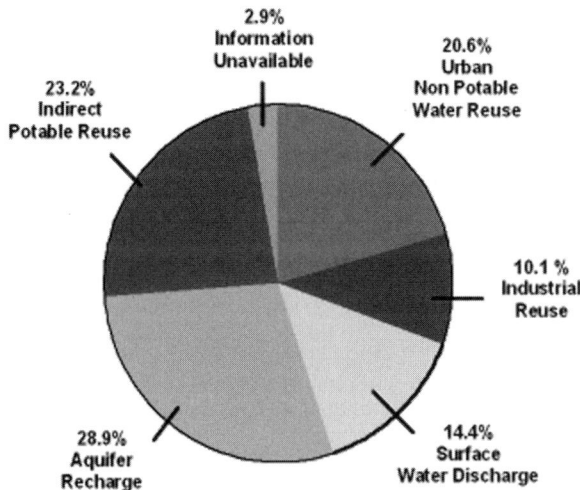

FIGURE 4.1—USE OF PRODUCT WATER FROM EXISTING FULL-SCALE WASTEWATER TREATMENT FACILITIES USING MEMBRANES.

## FEED WATER QUALITY

Feed water quality for low-pressure membranes can vary significantly depending on the type and efficiency of upstream wastewater treatment processes, the time of year (season), the amount of recent service area rainfall, and wastewater composition (e.g., percentage of domestic and commercial/industrial wastewater). Most low-pressure membranes receive secondary effluent from the secondary clarifiers following a biological treatment process, typically activated sludge. Table 4.1 presents a representative range of water-quality values for secondary effluent from municipal wastewater facilities that might reasonably become the feed

Table 4.1 Typical Feed Water (Secondary Effluent) Quality (Metcalf and Eddy, 2003; Reardon et al., 2005; WEF, 1992, 1996).

| PARAMETER | UNITS | SECONDARY EFFLUENT | REPORTED VALUES FOR MF FEEDWATER |
|---|---|---|---|
| Biochemical oxygen demand | mg/L[b] | 5–30 | 5–30 |
| Total organic carbon | mg/L as C | 5–30 | 5–30 |
| Total Kjeldahl nitrogen | mg/L as N | 10–20 | 5–30 |
| Total phosphorus | mg/L as P | 1–8 | 1–8 |
| Iron | mg/L | 0–1 | 0–1 |
| Total suspended solids | mg/L | 5–30 | 2–30 |
| Turbidity | NTU | 2–10 | 1–10 |
| pH | s.u. | 6–9 | |
| Temperature | °C | 8–30 | |
| Total coliforms[a] | No./100 mL | 50 000–2 000 000 | |
| Fecal coliforms[a] | No./100 mL | 10 000–1 600 000 | |
| Viruses[a] | PFU/100 mL | 0.05–1000 | |
| Protozoan cysts | No./100 mL | 5–10 | |
| Total residual chlorine | mg/L | 0–2 | |
| Fats, oils, and grease | mg/L | 0.1–10 | |

[a] Before disinfection.
[b] mg/L = ppm.

water for MF and UF facilities as well as a representative range of feed water quality values reported for MF facilities treating secondary effluent. At a minimum, secondary effluent must meet the minimum standards of the Clean Water Act, which require a neutral pH (6.0 to 9.0), a maximum monthly average of 30 mg/L (30 ppm) for five-day biochemical oxygen demand ($BOD_5$) and total suspended solids (TSS), and a maximum monthly geometric mean of 200/100 mL (200/12.2 cu in.) for fecal coliforms. Well-operated facilities designed to modern standards will provide a higher-quality effluent than is required by the Clean Water Act. Plants designed with advanced and tertiary treatment processes will provide further removal of conventional pollutants such as $BOD_5$ and TSS as well as nutrients. Particulate constituents and soluble compounds that can affect membrane performance are the feed water parameters of greatest interest in low-pressure membrane applications. Suspended; colloidal; and, to a lesser degree, dissolved matter may affect membrane performance.

## FILTRATE WATER QUALITY

Filtrate from MF and UF treatment facilities is generally characterized by consistent water quality, regardless of variations in the feed water composition. Microfiltration and UF membranes typically reject all suspended solids, provide 3 to 6 log removals of bacteria, reduce $BOD_5$ by at least 95%, and greatly reduce turbidity. Other contaminants such as phosphorus, nitrogen, and total organic carbon are also rejected by these membranes if the membrane process is combined with chemical (e.g., coagulation) or biological treatment. However, the rejection of these contaminants can cover a wide range, from 10 to 85%, depending on the phase (soluble or particulate) of the contaminant and the effectiveness of the chemical treatment. As with conventional tertiary treatment processes, the use of chemical coagulation with low-pressure membranes will improve the removal of soluble and colloidal constituents to the extent they can be coagulated or precipitated. Removal of particulates by low-pressure membranes depends on the pore size of the membrane relative to the particle size. Therefore, UF membranes should provide better removal of contaminants such as viruses that are smaller than most MF pore sizes but larger than the pores in a UF membrane. MF membranes can provide 3 to 6 log removal of protozoan cysts and coliform bacteria, while an intact UF membrane will provide complete removal.

Table 4.2 presents a representative range of filtrate water quality achieved by MF or UF treatment. It should be noted that upsets in upstream wastewater treatment processes can significantly affect MF and UF performance and can degrade filtrate water quality.

## TREATMENT MECHANISM

The mechanism by which MF and UF membranes operate is simple. Membranes are a physical barrier to suspended particles contained in the feed stream. All particles larger than the pore plus some fraction of particles that are smaller than the pore are retained on the feed side of

Table 4.2  Typical Filtrate Water Quality for MF and UF Treatment Facilities.

| PARAMETER | UNITS | RANGE OF VALUES | REMOVAL |
| --- | --- | --- | --- |
| Biochemical oxygen demand | mg/L[a] | <2–5 | 85–95%[b] |
| Total organic carbon | mg/L as C | 5–25 | 5–60%[b] |
| Total Kjeldahl nitrogen | mg/L as N | 5–30 | 6–8%[b] |
| Total phosphorus | mg/L as P | 0.1–8 | 1.5–3%[b] |
| Iron | mg/L | 0–0.2 | 0–20%[b] |
| Total suspended solids | mg/L | Below detection limits | >99% |
| Turbidity | NTU | <0.1 | 95–>99% |
| Silt density index | | <2–3 | NA[c] |
| Fecal coliforms | No./100 mL | <2–10 | 3 to >6 log |
| Virus | PFU/100 mL | <1–300 | 0.5 to 6 log |
| Protozoan cysts | No./100 mL | 0–1 | 3 to >6 log |

[a] mg/L = ppm.
[b] No removal can be expected in the absence of chemical treatment if the parameter exists only in soluble form; however, in secondary effluent, a significant fraction of this parameter is associated with suspended and colloidal solids.
[c] NA = not applicable.

the membrane. Additional rejection will be provided once a foulant layer accumulates on the membrane surface; however, periodic cleaning of membranes results in an inconsistent increase in rejection. Because these membranes reject suspended material based on size, the term "size exclusion membranes" has been used to describe MF and UF membranes. Neither MF nor UF membranes are capable of removing dissolved materials from the feed stream. However, some UF membranes are capable of removing high-molecular- weight compounds (approximately 100 000 to 300 000 Da for membranes in common use). Table 4.3 lists sizes for

Table 4.3 Apparent Dimensions of Small Particles, Molecules and Ions (Courtesy of Plenum Press).

| Species | Range of Dimensions (nm)[a] | | Molecular Weight (Da) |
|---|---|---|---|
| Secondary effluent TSS[b] | 1000 | 150 000 | |
| *Giardia lamblia* cysts | 8000 | 12 000 | |
| *Cyclospora cayetanensis* cysts | 8000 | 10 000 | |
| *Cryptosporidium parvum* cysts | 4000 | 6 000 | |
| Yeasts and fungi | 1000 | 10 000 | |
| Bacteria | 300 | 10 000 | |
| *Escherichia coli* | 1100 | 1 500 | |
| Oil emulsions | 100 | 10 000 | |
| Colloidal solids | 100 | 1 000 | |
| Viruses | 30 | 300 | |
| Hepatitis A virus | 27 | 27 | |
| Enterovirus | 25 | 30 | |
| Proteins/polysaccharides | 2 | 10 | $10^4$–$10^6$ |
| Enzymes | 2 | 5 | $10^4$–$10^5$ |
| Common antibiotics | 0.6 | 1.2 | 300–1000 |
| Organic molecules | 0.3 | 0.8 | 30–500 |
| Inorganic ions | 0.2 | 0.4 | 10–100 |
| Water | 0.2 | 0.2 | 18 |

[a] nm = nanometer; 1000 nm = 1 micrometer (micron; μm) = $4.0 \times 10^{-5}$ in.
[b] TSS = total suspended solids.

many pollutants to illustrate contaminants that can be removed by size exclusion membranes. For example, a MF membrane with a pore size of 0.2 μm ($10^{-5}$ in.) should remove suspended solids, protozoan cysts, yeasts, and some bacteria. Smaller particles, including viruses and ions, will pass through the membrane. Most membrane manufacturers publish a nominal pore-size diameter that can be used to estimate particle rejection; however, the nominal pore size can be misleading because membrane pores cover a range of sizes. For example, a MF membrane with a nominal pore size of 0.2 μm ($10^{-5}$ in.) might have pore sizes up to 5 μm ($2 \times 10^{-4}$ in.).

With time, the retained material accumulates on the surface of the membrane and increases the resistance to water flow through it. As a result of this fouling, the MF or UF processes must be periodically stopped for backwashing to recover a portion of the productivity lost through operation. Less frequently, they must be stopped and chemically cleaned to remove foulants accumulated on the surface of the membrane that were not removed by backwashing.

## *SAFETY*

Operator safety is of paramount importance in all treatment facilities; however, it is beyond the scope of this publication to address the safety aspects related to all available membrane equipment and all chemicals used in membrane processes. Therefore, it is recommended that workers thoroughly read and review all material provided by the membrane equipment manufacturer and follow all safety instructions. Material safety and data sheets for each chemical used should also be obtained and read and all safety instructions should be followed.

# Process Configurations

Low-pressure membrane effluent filtration systems typically consist of the MF or UF membrane system and various pretreatment and posttreatment systems. An example of a typical process flow schematic is shown in Figure 4.2. For the purposes of this chapter, pretreatment

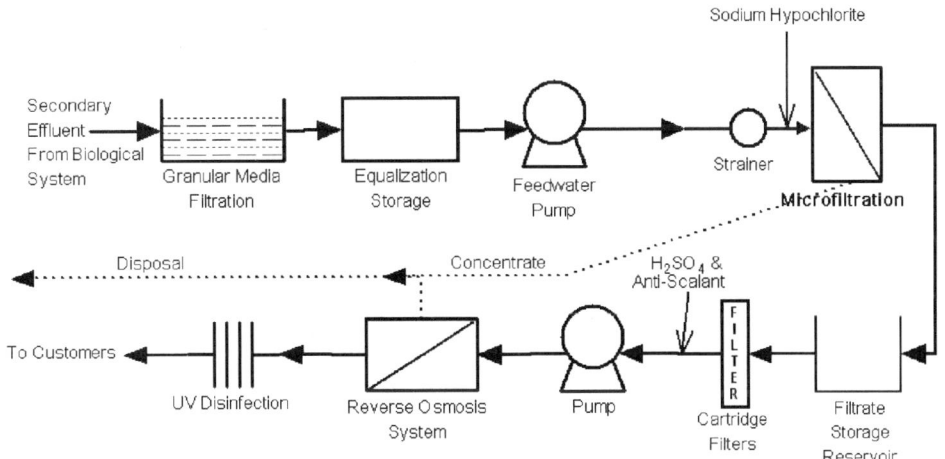

FIGURE 4.2—PROCESS FLOW SCHEMATIC FOR EFFLUENT MEMBRANE FILTRATION.

processes include all processes after the secondary clarifiers and before membrane filtration. Design and operation of each of these systems requires a thorough understanding of feed water quality, desired filtrate water quality, and capabilities of the treatment system. The configuration of typical pretreatment, membrane treatment, chemical storage and feed systems, posttreatment, and residuals handling processes are presented in this section.

## *PRETREATMENT*

Examples of common pretreatment processes include flow equalization, fine screens, strainers (Figure 4.3), chlorination, dechlorination, and chemical coagulation. However, the inclusion of these processes in any given treatment facility depends on influent water quality, filtrate quality requirements, and the membranes used. While in some instances secondary effluent can be fed directly to tertiary low-pressure membrane systems, experience has shown that pretreatment of the secondary effluent is essential to optimize operation of the membrane system. Several existing full-scale tertiary MF systems are preceded by both

**FIGURE 4.3—PRETREATMENT STRAINERS IN AN EFFLUENT MEMBRANE FILTRATION SYSTEM.**

granular media filtration and strainers. Prechlorination with a free or combined residual is desirable to minimize biological fouling, provided the membranes can tolerate the type and dose of residual applied. Dechlorination may also be required if MF or UF treatment follows chlorine or chloramine disinfection and membranes are not chlorine- or chloramines-tolerant. Dechlorination is typically accomplished via the addition of a strong reducing agent, such as sulfur dioxide or sodium bisulfite.

More elaborate pretreatment processes such as disinfection with processes other than chlorination, advanced oxidation, and inline coagulation may also be used. However, the need to implement these processes depends on feed water quality and filtrate water quality requirements. Besides chlorination and chloramination, disinfection may be accomplished with UV radiation, ozonation, or hydrogen peroxide. Inline coagulation may be used to improve rejection capabilities or backwashing efficiency; however, bench or pilot testing is required before implementation. Membrane compatibility must be verified before addition of any chemical upstream of the membranes.

## MEMBRANE SYSTEM

Membrane systems include membrane units and associated equipment required to maintain productivity, including air scour systems, backwashing or chemically enhanced backwashing systems, instrumentation, controls, and chemical cleaning systems. Figure 4.4 is a typical flow schematic that shows how membrane units and support systems are interrelated in an effluent filtration application. There are two types of membrane configurations: pressurized and immersed. Pressurized membrane configurations consist of membranes located within individual pressure vessels, with groupings of these pressure vessels housed in frames within buildings or on concrete pads. Immersed membrane configurations consist of membranes assembled into filter cells (also known

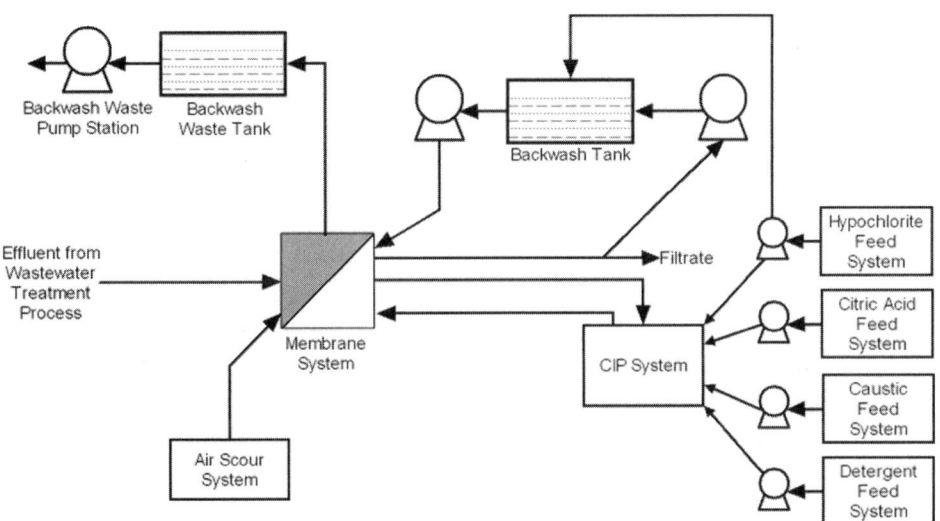

FIGURE 4.4—SCHEMATIC DIAGRAM OF A TYPICAL EFFLUENT MEMBRANE FILTRATION SYSTEM.

as racks or cassettes) located within one or more tanks containing the wastewater to be treated. Ancillary systems for both configurations are typically located adjacent to the tanks or pressure vessels.

## CHEMICAL STORAGE AND FEED SYSTEMS

The number of chemical storage and feed systems required for a membrane treatment facility is directly related to the number of chemicals added during the treatment process and backwashing and cleaning procedures. Common chemicals introduced to the feed water for MF and UF membrane systems include chlorine or sodium hypochlorite, hydrogen peroxide, ammonia, sodium bisulfite, sodium metabisulfite, ferric chloride, and/or alum. Chemicals most often used for cleaning are chlorine, citric acid, caustics, and/or detergents. Typically, one chemical storage and feed system is required for each chemical used at the membrane treatment facility. Chemical storage and feed systems typically consist of bulk chemical storage, a day tank (optional), chemical metering pumps, associated piping and valves, and mixers or diffusers. Bulk chemical storage is typically sized to accommodate a minimum 14- to 30-day supply. However, bulk storage for chemicals that degrade/decompose with time, such as sodium hypochlorite, are often sized smaller.

Chemical storage and feed systems may be manually controlled or fully automated. Manual systems should have, at a minimum, an indicator of the mass or volume of chemical in storage (visual tank level indicator or tank scale) and an indicator of the amount of chemical injected to the treatment process (flow meter). Fully automated systems will have this information incorporated to the control system. Furthermore, the chemical feed system will typically be flow-paced or controlled by a separate feedback or feed-forward system.

## POSTTREATMENT

Posttreatment processes required for MF/UF membrane treatment facilities are highly dependent on the end use of the filtrate. Disinfection, filtrate storage, and pumping are typically the only posttreatment systems required. However, some additional treatment processes may be used regionally to remove specific chemical compounds. One example is the use of hydrogen peroxide and UV to destroy N-nitrosodimethylamine.

# RESIDUALS HANDLING

Disposal of concentrated waste streams produced in membrane treatment processes is a key component of successful and sustainable operation of any membrane facility. Waste streams generated by MF and UF processes typically consist of spent cleaning solutions, backwash water, and concentrate streams.

Because low-pressure membranes do not remove soluble inert constituents, the backwash water is easily treated by recycling it back to the upstream biological treatment facility. For satellite low-pressure systems, the residuals can typically be discharged to the closest wastewater collection system. Other disposal or reuse options may be available, depending on local regulations and needs. Depending on the volume of water generated per backwash and the backwash frequency, the total backwash volume from most low-pressure membrane systems ranges from 2 to 10% of the daily feed water volume.

Spent cleaning solution volumes are small relative to the system capacity and are typically composed of solutions of acids, caustics, and detergents. As with the spent backwash, the spent cleaning solutions can typically be returned to the upstream biological treatment facility for processing. Care must be taken to ensure that pH extremes and/or chlorine residuals are not recycled back to the head of the biological process, which would cause system upsets. Typical characteristics for low-pressure membrane cleaning solutions are summarized in Table 4.4.

Should a satellite facility not have easy access to a wastewater treatment facility, the cost and complexity of waste stream disposal could become significant. Ease of permitting varies with the waste stream water quality and quantity; receiving water quality and quantity; preferred method of disposal; and local, state, and federal regulations. Alternative disposal methods might include land application, surface water discharge, and deep well injection.

Table 4.4 Typical Characteristics of Chemical Cleaning Solutions in Low-Pressure Membranes (Adapted from AWWA, 2003; WERF, 2005).

| CHARACTERISTIC | TYPICAL VALUE OR ATTRIBUTE[a] |
|---|---|
| Frequency of cleaning | Daily to once every three to four months |
| Volume of waste produced | Monthly clean-in-place wastes typically <0.05% of plant feed flow (daily chemically enhanced backwash wastes might be 0.2 to 0.4% of plant feed flowrate) |
| Chemicals commonly used | ■ Sodium hypochlorite: 500 to 1000 mg/L[b] as $Cl_2$<br>■ Citric or hydrochloric acid: 100 to 200 mg/L pH 1 to 2<br>■ Caustic soda: pH 12 to 13<br>■ Surfactant: 0.1% by weight |
| Characteristics of spent cleaning solution | ■ pH from 2 to 14<br>■ Chlorine residual up to 1000 mg/L as $Cl_2$<br>■ Low concentrations of surfactants<br>■ TSS up to 500 mg/L (neutralization may precipitate additional solids)<br>■ TOC 10 to 30 times the feed water concentration<br>■ $BOD_5$ up to 5000 to 10 000 mg/L if citric acid used |

[a] TSS = total suspended solids; TOC = total organic carbon; $BOD_5$ = five-day biochemical oxygen demand.
[b] 1.0 mg/L = 1.0 ppm.

# Equipment Configurations

This section describes the typical configuration of membrane equipment components as well as typical instrumentation and controls.

## PRETREATMENT COMPONENTS

Though nearly all wastewater treatment facilities include some type of mechanical screening as part of their preliminary treatment processes, no standard practice exists regarding the type of screen or aperture sizes to be used. As a result of variable screening capabilities at the headworks of wastewater treatment plants, the capture of windblown debris in open tanks and the inability of most existing solids-separation processes to remove neutrally buoyant materials, solids capable of clogging a low-pressure membrane system, will be present in most secondary effluents. In addition, significant amounts of algae are often present in secondary effluent, at least on a periodic or seasonal basis. Secondary effluent from attached growth processes can also contain significant quantities of small organisms such as snails, flying insects, and worms. At a minimum, pretreatment for MF and UF systems used for effluent filtration should consist of fine screening or strainers (200 to 2000 μm, or 0.008 to 0.08 in.) to remove large suspended material (large relative to the operation of low-pressure membranes). Table 4.5 summarizes typical opening sizes for common categories of screens and strainers used for membrane pretreatment.

Immersed membrane systems require the least stringent solids removal from the feed water, with aperture sizes of 1 to 3 mm (0.04 to 0.12 in.) typically being sufficient. Pressurized, outside-in membranes typically require finer solids removal, typically with an aperture opening of 1 mm (0.04 in.) or less. Pressurized inside-out membranes need the best pretreatment for solids removal, with aperture openings of approximately 200 μm (0.008 in.) required.

Table 4.5 Typical Opening/Pore Sizes for MF/UF Membranes and Common Pretreatment Devices.

| Membrane | Typical Range of Opening Sizes (μm)* | | Molecular Weight Cutoff (Da) |
|---|---|---|---|
| Fine screen | 1500 | 6000 | NA |
| Granular media filtration | | | NA |
| Media effective size | 350 | 2000 | |
| Smallest particle removed | 8.0 | 12.0 | |
| Prefilter or strainer | 200 | 500 | NA |
| Microscreen | 1 | 200 | NA |
| Microfiltration | 0.1 | 0.2 | NA |
| Ultrafiltration | 0.01 | 0.1 | 100 000–300 000 |

* μm = micrometer (micron); 1000 μm = 1 mm = 0.04 in.

Screens or strainers used for membrane treatment should be designed for severe duty applications. They should also be designed for continuous or automatic backwashing based on a pressure differential or a time interval. Manual screens, especially units with openings smaller than 1000 μm (0.04 in.), are not typically practical because they require almost constant cleaning to avoid a hydraulic bottleneck and limit flow to the downstream membrane system.

No consensus currently exists on the need for granular media filtration (GMF) before low-pressure-membrane treatment of secondary effluent. Granular media filtration, especially when combined with prechlorination and chemical coagulation, will significantly reduce the mass load of foulants applied to the membrane system and should extend membrane cleaning intervals or allow the membranes to operate at a higher flux. While GMF as a pretreatment step for low-pressure membranes is generally helpful from an operational perspective, pilot-plant testing and economic evaluations are required before deciding on its necessity or feasibility. Granular media filtration may provide insufficient benefit to justify the cost.

## MEMBRANE EQUIPMENT COMPONENTS

Components used in typical MF and UF treatment facilities include membranes and associated hardware, a permeation system, a flux maintenance system, a clean-in-place (CIP) system, an integrity testing or monitoring system, and a programmable logic controller and associated instrumentation. Systems used for wastewater effluent filtration can be either immersed or pressurized. Figure 4.5 shows an example of an immersed low-pressure (MF/UF) membrane system. Figure 4.6 is a photograph of a pressurized low-pressure (MF/UF) system. Details about these systems are covered in Chapter 2.

# Operation

This section describes the operation of MF and UF membrane treatment facilities filtering secondary effluent from a wastewater treatment facility. An operational overview of pretreatment processes, automated

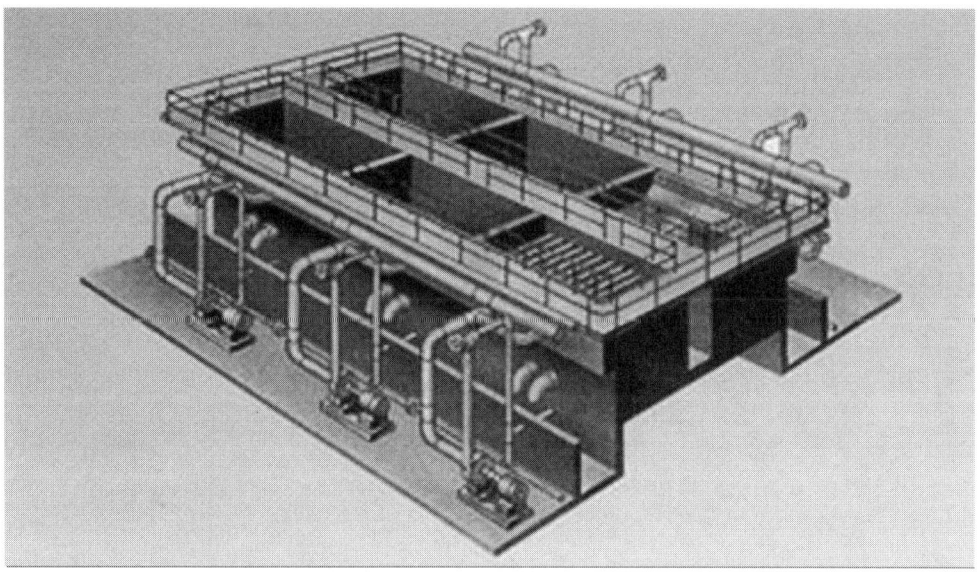

FIGURE 4.5—ILLUSTRATION OF AN IMMERSED LOW-PRESSURE MEMBRANE SYSTEM (COURTESY OF USFILTER).

systems, and event-triggered systems will be presented. In addition, capacity management and process optimization will also be addressed.

## PRETREATMENT

Most pretreatment for MF/UF facilities treating secondary effluent occurs in the secondary treatment processes. Therefore, it is imperative that all secondary treatment processes are operated properly. However, it is beyond the scope of this publication to discuss the operation of conventional activated sludge (CAS) processes. Additional operational information on CAS systems can be found in several of the references listed in Appendix II of this publication. There are, however, some unique operational considerations that should be taken into account for an activated sludge treatment process that is followed by a MF/UF membrane system. For example, changing the sludge age or eliminating/changing the polymer may be necessary to minimize membrane fouling. Operating conditions such as short solids retention times or nutrient deficiencies that can result in high concentrations of exocellular

FIGURE 4.6—PRESSURIZED MF TREATMENT SYSTEM (COURTESY OF USFILTER).

polymers in the biomass should be avoided, if possible. High overflow rates on secondary clarifiers can result in increased effluent suspended solids, which, in turn, increase solids loading on membrane systems. However, often the secondary wastewater treatment facility already exists, and the MF/UF retrofit design and operations plan must take into account the variable nature of biologically treated secondary effluent. Periodic cleaning of clarifier weirs and launders can result in slug loads of algae in the secondary effluent, and shutdown of the membrane system during cleaning should be considered if the membrane pretreatment cannot handle large amounts of algae.

Even if the membrane process is immediately preceded by fine screens or strainers, adequate fine screening with redundancy should still be provided before the activated sludge treatment process. For more information on fine screens located at the headworks of a treatment facility that includes membranes, please refer back to Chapter 2.

Following secondary treatment, fine-screening units or strainers are typically used as a pretreatment process for low-pressure membrane facilities. Fine screening units with automatic or continuous self-cleaning are often installed at MF or UF facilities, as they are easy to operate and require little maintenance. Self-cleaning strainers remove suspended material larger than the screen size from the feed water. Operation of the strainer will depend on the type provided. Some strainers continuously backwash, whereas others only backwash after a predetermined pressure drop or time interval. Once this pressure drop is reached, a flushing valve is opened and the strainer is backwashed. After the backwash is completed, the strainer automatically returns to service. Manually operated screens or strainers are not recommended. Figure 4.7 illustrates operation of a typical strainer.

Operation of pretreatment processes requiring the injection of chemicals, such as dechlorination, disinfection, and inline coagulation, are very similar. The injection of chemicals to the treatment system is accomplished by a system that typically consists of bulk chemical storage, an optional day tank, chemical metering pumps, and a calibration column. Care must be taken to provide adequate mixing of chemicals with the feed water. Chemical metering systems may be manually operated or fully automated. Manual operation of chemical injection systems typically consists of frequently checking the feed water flow rate and adjusting the chemical metering rate accordingly. All chemical metering pumps should be periodically calibrated using the calibration column and adjusting settings as required. Automated chemical injection systems should be flow-controlled at a minimum so the chemical feed rate automatically increases or decreases with plant flow; however, automatic flow control is not necessary for membrane systems operating at a constant rate. Automatic dosage control is seldom warranted unless there are significant and frequent changes in the required chemical dose. Both manual and automated chemical injection systems require periodic refilling of bulk storage and optional day tanks.

## *AUTOMATED SYSTEMS*

By their very nature, membrane facilities require a high degree of automation. Automated systems allow for continuous monitoring and automatic

**132** ■ *Membrane Systems for Wastewater Treatment*

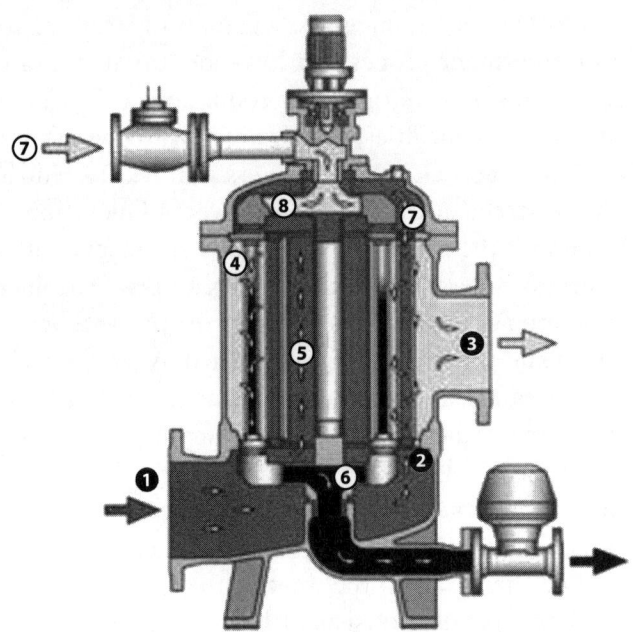

**Straining Process:**
① Inlet flange.
② Dirty fluid flows in through both ends of the elements.
③ Outlet flange.

**Automatic Cleaning Cycle:**
④ Backflush fluid flows from the throttled top into the elements.
⑤ Clean fluid flows in from the outside of the elements.
⑥ Strained particles are flushed down the sludge drain valve.
⑦ Introduce external flush meduim of L.P. steam or high pressure filtered water.
⑧ Steam or water assist in the cleaning process.

**FIGURE 4.7—CUTAWAY DIAGRAM OF A TYPICAL STRAINER (COURTESY OF USFILTER).**

control of a variety of components. Typical automated systems include filtrate flow control, transmembrane pressure (TMP) alarms, and backwash systems. Each of these systems is briefly discussed below.

Filtrate Flow Control

On vacuum-driven systems, filtrate flow controls automatically monitor and adjust the filtrate flow for a group of membranes. In most systems, a flow-control device (feed pump, filtrate pump, or filtrate valve) is controlled by the output of a filtrate flow meter. However, immersed systems may use the output of a membrane tank liquid level indicator to control the flowrate. The number of filtrate flow control devices installed at a membrane facility depends on the size of the facility, organization of the membranes, and level of control desired. Typically, one filtrate flow control system is installed on each membrane skid.

Transmembrane Pressure Alarms

Transmembrane pressure alarms automatically monitor the TMP for a group of membranes. An alarm is activated if the differential pressure across the membranes exceeds a predetermined setpoint, which is typically specified by the membrane manufacturer.

Figure 4.8 contains a graph of feed, filtrate, and TMPs for low-pressure membrane systems treating secondary effluent. This provides a good example of TMP variations during normal operation of a low-pressure membrane system.

Backwash System

The backwash system provides for the automatic initiation of a backwashing sequence and controls the backwash pumps and blowers (where applicable); the feed or filtrate pumps; and associated automatic valves on the feed, filtrate, and drain lines. Details of the general operation of these systems can be found in Chapter 2.

# PRODUCTIVITY MANAGEMENT

Low-pressure membranes can be operated to maintain a constant filtrate flowrate (constant flux) or to maintain a constant TMP (constant pressure); however, full-scale production systems are always designed to operate in a constant flux mode. Severe fouling combined with limited feed water pressure may result in operation in a declining flux mode; however, this is an upset condition. As with many facilities, maintenance of treatment capacity is essential for reuse facilities that are supplying water for critical applications such as industrial process and cooling water. Prudent design practices will provide product water storage and some level

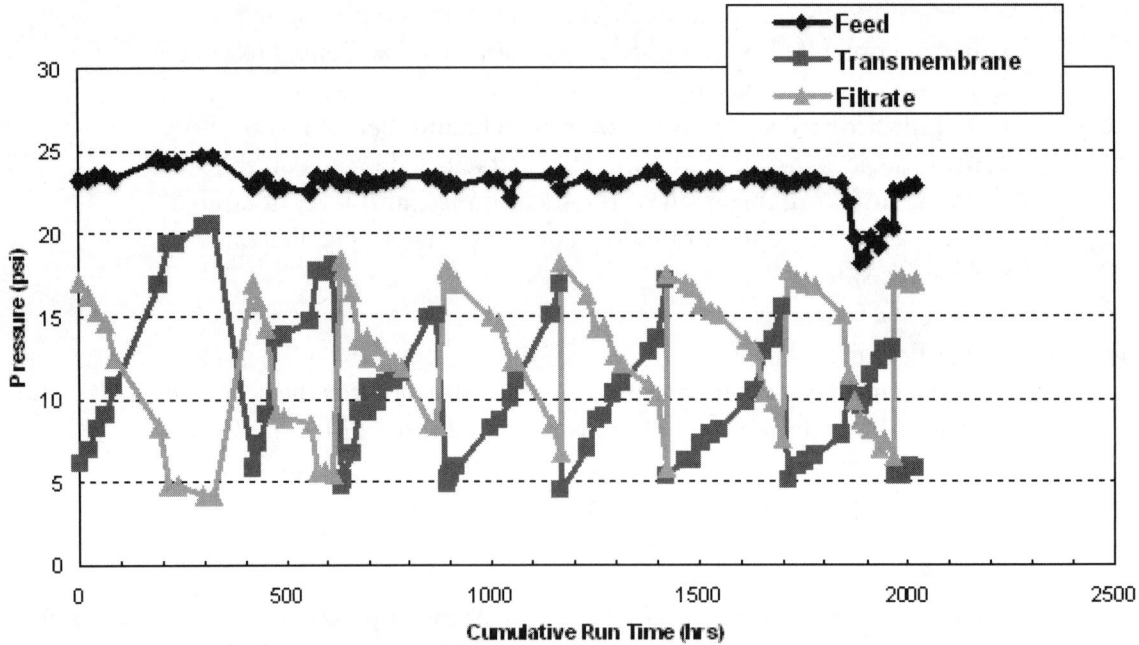

FIGURE 4.8—TYPICAL OPERATING PRESSURES FOR LOW-PRESSURE MEMBRANES FILTERING SECONDARY EFFLUENT (psi × 6.89 = kPa) (ADAPTED FROM WERF, 2005).

of redundancy for key equipment by including online spares. This enables the operator to maintain full treatment capacity during equipment malfunctions. However, scheduling and conducting routine maintenance procedures are at the discretion of the operator. Care should be taken to schedule and conduct maintenance procedures during periods of low flow or at times when it is safe to shut down a process train (if required).

Optimization of low-pressure membrane processes to maximize or at least maintain production capacity while minimizing operating costs can be achieved through the adjustment of several key operating parameters, including pretreatment, applied flux, and backwashing and CIP procedures and frequency. Each of these parameters will be discussed below.

Pretreatment

The life and performance of low-pressure membranes can be maximized through the proper use of physical and chemical pretreatment processes. Optimizing pretreatment processes helps to maintain feed water flow by minimizing membrane fouling and the frequency of backwashing and chemical cleanings. Strainer operation should be monitored to ensure that pressure loss across the strainer does not increase beyond design values and reduce feed flow to the membrane process. Fouling of MF or UF membranes can be reduced by chemical coagulation of the feed water. Low doses (<5 mg/L, or <5 ppm) of ferric chloride are often used with inside-out UF membranes to control membrane fouling. When the low-pressure membrane is being used to remove phosphorus or another soluble constituent, proper control of the chemical dose is essential to obtain the desired removal without putting excessive solids loading on the membranes. Jar testing is a useful tool for evaluating the potential effect of changes in chemical dose to affect performance; although jar testing cannot simulate actual changes to membrane performance. However, as with inline filtration, the settle ability of floc is not a controlling factor and suspended micron-sized floc may be sufficient.

Design Operating Flux

Generalizations about operating fluxes to be expected with low-pressure membranes treating secondary effluent are difficult. With current membranes and systems, pilot testing is still essential for establishing the design flux for each facility. Design operating flux varies with the fouling characteristics of the secondary effluent, membrane material, and membrane system configuration. Immersed membranes treating secondary effluent from a relatively high-rate activated sludge process will likely operate in the range of 27 to 41 $L/m^2 \cdot h$ (16 to 24 gpd/sq ft), while pressurized inside-out systems may be able to sustain rates of 51 to 59 $L/m^2 \cdot h$ (30 to 35 gpd/sq ft) on low-fouling secondary effluent.

Operation of a low-pressure membrane facility at elevated flux values can result in increased rates of fouling. Once the flux increases above certain values, the rate of fouling can increase exponentially. Pre-design pilot testing and routine monitoring of system operation are needed to be certain that the membrane flux is below the value at which deterioration of system performance begins to accelerate. Operation of a MF or UF membrane facility at reduced flux values will reduce energy consumption but can result in the inefficient use of installed membrane capacity, and the design capacity of the system may not be achieved. The design operating flux should be optimized to maintain a balance between acceptable fouling rates, operating costs, and economical use of membrane capacity. Though the operating flux for a low-pressure membrane process has a significant effect on the rate of fouling, operators have few options for controlling flux. When demand is less than design capacity, flux can be reduced by maintaining all units in operation. However, if demand is at design capacity and the design flux results in unacceptable fouling despite optimization of pretreatment and backwash procedures, limited choices are available. Backwashing and cleaning frequencies must be increased, and production must be reduced.

### Backwashing Frequency

Backwashing removes the layer of contaminants retained on the feed side of the membrane that have accumulated during the previous operating cycle. This fouling layer presents additional hydraulic resistance to fluid flow across the membrane. To overcome this additional resistance, elevated TMPs are required, which results in increased operating costs. The frequency of backwashing events should be optimized to maintain low TMPs throughout normal operating cycles. Because filtrate production stops when a unit is backwashed, increasing the frequency or duration of backwashes reduces the net daily production of water. Further, most systems require filtrate for all or a portion of the backwash water so that use of increased volumes of backwash water reduces the overall recovery of water. When the total daily backwash time and volume exceed the design values, overall system production will drop below design.

### Clean-In-Place Sequence

Chemical cleaning removes contaminants not removed during normal backwashing. The frequency of chemical cleaning should be selected to maintain low TMPs throughout the life of the membranes. Existing low-pressure membrane facilities treating secondary effluent have reported chemical cleanings at intervals as infrequent as once every 2 months to as frequent as once every few days. Depending on the fouling characteristics of an individual wastewater and the use of chlorine before the membranes, a cleaning frequency of once every 2 to 4 weeks should be expected. Consideration should be given to the design of a chemical neutralization system.

Seasonal Changes in Operation

Biological processes producing feed water for low-pressure membranes should be optimized during seasonal changes such that a consistent effluent is produced throughout the year. Should upsets occur in the secondary treatment processes, the secondary effluent may contain elevated concentrations of suspended solids and organic material, which can have adverse effects on a subsequent membrane process. Potential adverse effects include elevated contaminant loading rates, increased fouling rates, reduced flux, elevated pressure requirements, increased backwashing frequency, increased CIP frequency, reduced backwashing efficiency, and reduced chemical cleaning efficiency.

# Routine Monitoring

Successful operation of any treatment process requires the frequent monitoring of significant parameters, and MF and UF processes are no exception. Monitoring parameters of interest for pretreatment and membrane systems are described below. In addition, a brief discussion on the importance of routine recording of data and calculating data trends is presented.

## PRETREATMENT SYSTEM

Monitoring parameters for pretreatment systems (fine screens, strainers, and chemical additions) include flow or volume, influent pressure, effluent pressure, turbidity, feed water TSS, and chemical metering rates (if applicable). All of these parameters, except suspended solids, should be monitored continuously. Although suspended solids probes are available, use of the suspended solids values measured in the secondary effluent for permit compliance is sufficient. Pressure instrumentation or level sensors should provide instantaneous pressure or level measurements for pretreatment influent and effluent streams to allow for determination of the pressure drops across pretreatment processes, including screens or strainers, inline mixers, and associated piping and valves. This enables the operator to determine if pretreatment processes are operating properly. Instrumentation associated with chemical metering rates should indicate and record chemical feed rates or volumes. Provisions to track weekly and monthly information are recommended for chemical feed rates.

## MEMBRANE SYSTEM

Feed water and filtrate flow (or volume), temperature, pH, feed side and filtrate side pressures, feed water and filtrate turbidity, and chemical metering rates should be monitored continuously. Though membrane permeability (flux/TMP), also referred to as the mass-transfer coefficient (MTC), is a parameter of significant interest, it is not measured directly and must be calculated from direct measurements of pressure and flow. Flow meters should be configured to provide instantaneous flow, daily average flow, daily maximum flow, daily minimum flow, and total daily volume for all streams associated with MF or UF operation. These streams include feed, filtrate, concentrate (if used), recycle (if used), backwash, chemical enhanced backwash, CIP, and flush. Provisions to track weekly and monthly information are also recommended. Pressure instrumentation should be configured to provide instantaneous pressure measurements for all streams associated with MF and UF operation. Instrumentation for measurement of turbidity and particle counts for feed, filtrate, and flush streams should provide instantaneous information feedback. This enables the operator to verify proper operation of the membrane treatment and cleaning processes.

## TEST PARAMETERS

Table 4.6 summarizes minimum testing requirements. Instrumentation and control systems provide operators with invaluable information regarding the operating status and performance of the MF or UF

Table 4.6 Minimum Testing Requirements for MF and UF Facilities.

| PARAMETER | MINIMUM TEST FREQUENCY | RECOMMENDED TEST FREQUENCY | SAMPLE LOCATION | METHOD[a,b] |
|---|---|---|---|---|
| Total suspended solids | Weekly | Daily | Feed/filtrate | 2540 |
| Temperature | Daily | Continuous | Feed | NA |
| pH | Daily | Continuous | Feed/filtrate | 4500-H |
| Turbidity | Weekly | Continuous | Feed/filtrate | 2130 |
| Particle count | Weekly | Continuous | Feed/filtrate | 2560 |
| Fecal coliform | Weekly | Daily | Feed/filtrate | 9222 |

[a] Methods are in *Standard Methods for the Examination of Water and Wastewater* (latest approved edition).
[b] NA = not applicable.

system; however, these systems are not capable of monitoring all parameters of interest. Furthermore, the output of these systems must be frequently verified, as they do not maintain calibration indefinitely. Therefore, routine sampling and testing are required.

## DATA ANALYSIS AND REPORTING

Even though many membrane systems have a supervisory control and data acquisition system, data should be manually recorded at least once per shift, and long-term trends updated and plotted daily. At a minimum, values should be recorded daily for filtrate flow, concentrate (or bleed) flow if used, recycle flow if used, influent and effluent pressure for pretreatment strainers or screens, temperature, pH, run times, membrane feed pressures (top and bottom if appropriate), filtrate pressure, and filtrate turbidity or particle counts. Any changes in operations or other notable events that have occurred such as errors in control, bad service, and upsets should also be recorded. Manually recording data encourages operators to physically inspect the facilities, verifies proper operation of electronic instrumentation, and provides a paper backup should electronic data be lost.

Trends in operating data can provide operators with valuable information regarding the status of their treatment facility. Flux and recovery should remain constant; any changes must be investigated and appropriate corrective action should be taken. Declines in membrane permeability (MTC) can be an indication of ineffective backwash procedures. Flow or volume trends will show if system productivity is declining over time. Downward flow trends might result from declines in pump performance or gradual increases in hydraulic losses. Transmembrane pressure is used in the calculation of permeability, as is temperature-corrected flux rate and area. Assuming a constant flux rate, temperature, and effective area, generalizations about system performance can be made from the TMP alone. Transmembrane pressure data provide an indication of the effectiveness of each backwashing sequence, as reductions should be observed after each occurrence. Long-term increases in the TMP may indicate a need to initiate a CIP sequence, while long-term decreases may indicate membrane degradation. Sudden decreases in TMP are a sign of membrane damage. Turbidity or particle count data provide an indication of the integrity of the membrane system. Filtrate turbidity should remain relatively constant with varying TMPs and fluxes, provided both are within the design ranges. Sudden increases in turbidity or particle counts may indicate that damage has occurred to the membrane, process piping, or process seals. Should an increase in turbidity or particle count be observed, the operator should perform an integrity test to identify any damaged membranes, process piping, or process seals.

Membrane performance is strongly affected by changes in temperature, as membrane permeability decreases with decreasing temperatures. Performance also decreases with increasing water viscosity, which changes inversely with water temperature. Therefore, it is important to note again that all data must be normalized to a standard water temperature, 25° C (77° F).

## Maintenance

Routine maintenance is required for any system to sustain optimum performance. Routine maintenance procedures for each membrane system will vary with membrane type, membrane materials, operating conditions, and influent water quality. As a result, it is recommended that the operator refer to the system-specific operations and maintenance manual from the manufacturer for a detailed maintenance schedule and procedures. However, the following list presents general maintenance procedures common to many membrane facilities. This is not a complete record of maintenance procedures and should not be used as a substitute for manufacturer-recommended procedures and frequencies.

## CHEMICAL CLEANINGS

Regular backwashing is very effective in removing a significant portion of contaminants retained on the feed side of the membrane; however, a fraction of these contaminants remains on the surface of or embedded in the membrane. Periodic chemical cleanings will be required to recover a portion of the productivity not recovered during normal backwashing sequences. Chemical cleaning procedures vary with membrane manufacturer, membrane configuration, membrane material, type of suspected foulant, and degree of fouling. Procedures for chemical cleanings range from a prolonged backwashing cycle enhanced by chemical addition to extended periods of immersion in a chemical bath. Chemicals typically used in chemical cleanings include acids, bases, and surfactants. Chlorine- and chloramines-resistant membranes may also be disinfected through the addition of free or combined chlorine residuals. Before addition of any chemicals to the membrane system, compatibility and recommended concentrations must be verified by the manufacturer.

## INTEGRITY TESTING

Daily integrity testing of low-pressure membranes is required for potable water treatment; however, no integrity testing has yet been required by wastewater permitting agencies. While the pressure-decay test is an automated feature of some systems, each membrane unit still must be taken out of production to conduct direct integrity tests. Thus, the frequency of direct integrity testing must balance the need to verify membrane integrity and the need to minimize system downtime. Table 4.7 lists approximate times to test individual mem-

Table 4.7  Typical Membrane Integrity Test Times (Courtesy of the Drinking Water Inspectorate, Department for Environment, Food and Rural Affairs).

| Membrane Integrity Test | Test Times (per module) |
|---|---|
| Diffused airflow test | 15 min |
| Pressure decay test | 10 min |
| Audible test–stethoscope | 1 min |
| Acoustic sensors | Direct |
| Molecular weight cutoff | 2 h |
| Challenge tests–microbiological | 24–48 h |
| Challenge tests–particle | 30 min |

brane modules. System size will determine the total testing time. Low-pressure membrane filtrate from the treatment of effluent will probably not be used directly for indirect potable reuse without subsequent treatment by RO. For this reason daily integrity testing is not necessary; however, if production capacity is adequate to afford the loss, automated systems should be tested daily.

Should an unacceptable breach of integrity be discovered during testing, the damaged block must be isolated until it is repaired or replaced. Typically, damaged hollow-fiber membranes may be repaired by plugging the damaged fiber with a metal pin or sealing the damaged area with an epoxy resin. Techniques used to repair size-exclusion membranes can vary by manufacturer, membrane material, type, and flow configuration. Before attempting to repair a damaged membrane, the manufacturer should be consulted for specific recommendations and instructions. Figure 4.9 shows examples of the pressure-decay rates from one broken fiber in several sizes of membrane modules as compared to pressure decay resulting from diffusion alone.

## TANK WASHING AND CLEANING

Process tanks that house immersed membrane systems should be periodically removed from service for cleaning, inspection, and maintenance. Some immersed systems completely drain the membrane tank

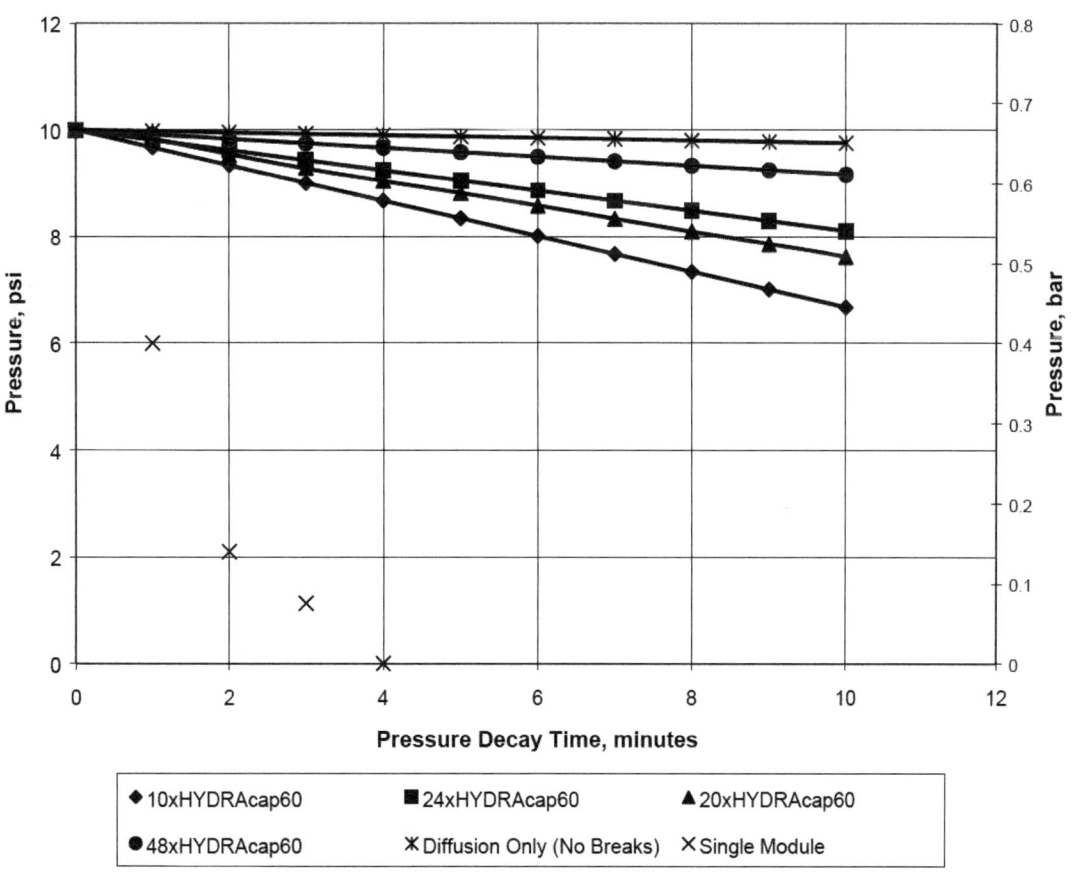

FIGURE 4.9—PRESSURE DECAY RATES IN PRESSURIZED UF FRAMES WITH ONE BROKEN FIBER AND WITH NO BREAKS (psi × 6.89 = kPa; bar × 100 = kPa).

after each backwash and, thereby, remove all accumulated solids; however, other systems do not regularly drain the tanks.

For the latter systems periodic cleanings should be performed to remove accumulated solids that were not completely removed during automated backwashing procedures. Because shutting down tanks for cleaning reduces the overall facility capacity, tanks should be removed from service during periods of low filtrate demand. To minimize the potential for membrane damage, manufacturers should be consulted so that the recommended procedures are followed. Figure 4.10 shows an example of the tank drain system for an immersed low-pressure membrane facility.

## INSTRUMENT CALIBRATION

Measurements obtained from a variety of instruments have a significant role in the operation of membrane treatment facilities. As such, the calibration of all instrumentation (flow meters, pressure gauges, turbidimeters, particle counters, etc.) should be verified on a regular basis. In the event that an instrument is determined to be out of calibration, the instrument should be recalibrated using known standards and the procedures provided by the manufacturer.

FIGURE 4.10—PHOTOGRAPH OF IMMERSED MEMBRANE TANK DRAINS (COURTESY OF ZENON ENVIRONMENTAL, INC.).

# Troubleshooting

Operation of any full-scale wastewater facility requires a staff of operators with a wide variety of troubleshooting skills. Variations in design, operation, equipment, and configuration preclude the generation of a complete list of troubleshooting guidelines in this book. However, general troubleshooting guidelines for low-pressure membrane systems that are filtering wastewater treatment plant effluent are presented in Table 4.8

Table 4.8  General Troubleshooting Guidelines for MF and UF Facilities (TDWI, 2001).

| Problem | Possible Cause | Corrective Action |
| --- | --- | --- |
| Decrease in feed flow to membrane unit | Damaged feed pump | Inspect feed pump for damage and repair/replace as required |
| | Closed valve | Check valve positions and adjust if necessary |
| | Clogged screen or strainer | Inspect screen or strainer and replace or clean as required |
| | Obstruction in feed piping | Inspect piping for obstructions and clean as required |
| Increase in feed stream turbidity/suspended solids | Improper operation of upstream wastewater treatment processes | Check upstream wastewater treatment processes for proper operation and adjust as required |
| | Improper operation of upstream screen or strainer | Check upstream screen or strainer for proper operation |
| Erroneous/nonsensical values obtained from instrumentation | Instrumentation out of calibration | Calibrate instrumentation with known standards per manufacturer's instructions |
| | Fouled instrumentation | Inspect instrumentation for fouling, clean instrumentation per manufacturer's instructions, and recalibrate per manufacturer's instructions |
| Sharp increase in TMP and no change in rejection | Degradation in feed water quality | Check upstream and pretreatment processes for proper operation |

Table 4.8  General Troubleshooting Guidelines for MF and UF Facilities (TDWI, 2001) *(continued)*.

| Problem | Possible Cause | Corrective Action |
|---|---|---|
| | Closed valve | Check valve positions and adjust if necessary |
| | Increase in flow | Check flow and adjust if necessary |
| | Clogged screen or strainer | Inspect screen or strainer and replace or clean as required |
| | Obstruction in piping | Inspect piping for obstructions and clean as required |
| Sharp decrease in TMP and sharp decrease in rejection | O-ring or seal failure | Inspect O-rings or seals for damage and repair/replace as required |
| | Membrane damage | Inspect membranes for damage and repair/replace as required |
| Gradual increase in TMP and no change in rejection | Biological or particulate fouling | Backwash or chemically clean the membranes as required |
| Gradual increase in TMP and gradual increase in rejection | Scaling (rarely occurs in low-pressure membranes) | Chemically clean the membranes as required |
| | | Add antiscalant to the feed stream |
| Gradual decrease in TMP and gradual decrease in rejection | Membrane damage | Repair or replace the membranes as required |
| | | Inspect membranes for damage and repair/replace as required |

## CHAPTER FIVE

# Nanofiltration and Reverse Osmosis for Advanced Posttreatment

Introduction .................................................. 149
    Applications .............................................. 149
    Water Quality ............................................. 150
    Treatment Mechanism ...................................... 151
    Equipment Manufacturers .................................. 152
    Safety ................................................... 152
Process Configurations ........................................ 152
    Pretreatment ............................................. 153
        Particulate Control .................................. 154
        Chlorination/Dechlorination .......................... 154
        Scale Prevention ..................................... 156
            Polyphosphonates ................................. 157
            Polyacrylates .................................... 158
    Membrane Treatment ....................................... 158
    Chemical Cleaning Systems ................................ 159
    Posttreatment Facilities ................................. 159
    Concentrate Disposal ..................................... 160
Equipment Configurations ...................................... 161
    Membrane Process Equipment Components .................... 162
        Two- and Three-Stage Membrane Array (Concentrate Staged) .. 162
        Two-Pass or Partial Two-Pass Membrane Array (Permeate Staged) ....... 163

Membrane Elements ............................................. 163
Typical Instrumentation and Controls ............................ 164
Operation ............................................................ 164
Pretreatment Chemistry .......................................... 165
Fouling ........................................................... 165
Sulfate Salts ................................................. 166
Silica ........................................................ 166
Organics ..................................................... 167
Instrumentation and Controls .................................... 167
Recovery ..................................................... 167
Feed and Differential Pressure ................................ 168
High and Low pH .............................................. 168
Event-Driven Systems ............................................ 168
Cleaning Events .............................................. 169
Source Water Changes ......................................... 170
Biofilm Development .......................................... 170
Capacity Management ............................................. 170
Process Optimization ............................................ 171
Routine Monitoring ................................................... 172
Feed Water ...................................................... 172
Membrane System ................................................. 174
Data Interpretation and Reporting ............................... 175
Percent Rejection ............................................ 176
Mass-Transfer Coefficient (Specific Flux) .................... 176
Differential Pressure ........................................ 176
Maintenance .......................................................... 177
Chemical Cleanings .............................................. 177
Integrity Testing ............................................... 178
Instrument Calibration .......................................... 180
Troubleshooting ...................................................... 181

# Introduction

This chapter will provide an overview of advanced posttreatment membrane wastewater systems: a description of current membrane process specifics for the use of reverse osmosis (RO) for wastewater treatment. Reverse osmosis is intended to represent both RO and nanofiltration (NF) technologies. Both technologies are pressure-driven membrane processes that use similar equipment and configurations. The primary differences between the two technologies are that RO membranes can remove monovalent ions (such as salts) and NF membranes allow monovalent ions to pass through. Because RO removes smaller particles, it requires a higher applied pressure than NF. Currently, there are no full-scale NF applications in advanced wastewater treatment. This section addresses configuration, operation, performance, monitoring and controls, maintenance, and troubleshooting of advanced posttreatment membrane systems.

## APPLICATIONS

During the last 20 years, the application of integrated membrane systems has emerged as a viable alternative for providing additional treatment to wastewater effluent for reclamation or reuse. This chapter focuses on RO applications to treat tertiary wastewater effluent (after treatment by conventional activated sludge /microfiltration [MF]/ultrafiltration [UF] or by membrane bioreactors [MBRs]) reuse or other reclamation uses. Limited water supply sources and increased regulatory requirements have resulted in many utilities assessing the option of wastewater reuse as an alternative source of water supply, aquifer recharge, or surface water recharge. In municipal wastewater treatment plants, RO membrane processes are used to provide advanced posttreatment to improve the quality of the effluent water for subsequent use. Permeate from RO membrane processes have reduced salts, disinfection byproducts, and emergent pollutants of concern (EPOCs) such as pharmaceuticals and hormones (Drewes et al., 2002). Therefore, RO membranes will continue to be used for municipal wastewater effluent posttreatment wherever the subsequent user or discharge mechanism requires superior quality water.

## WATER QUALITY

Feed water quality for RO membrane wastewater applications can vary significantly, depending on the local wastewater composition and upstream wastewater treatment processes. This subsection identifies typical feed water quality when the RO system follows a MBR or a biological secondary treatment system with MF/UF polishing and the typical product water quality expectations for both scenarios. Table 5.1 presents a representative range of feed water quality values for frequently monitored wastewater parameters at municipal wastewater facilities that use MBRs or MF/UF facilities upstream of RO facilities. Feed water qualities listed in Table 5.1 are taken from projected water qualities expected downstream of MBR and MF/UF

Table 5.1 Typical Feedwater Quality Ranges for RO Treatment Facilities.

| Parameter | Units | Range of Values: MBR as Pretreatment | Range of Values: CAS[a] Followed by MF/UF as Pretreatment |
|---|---|---|---|
| Total dissolved solids | mg/L[b] | 500–1500 | 500–1500 |
| Bicarbonate | mg/L as $CaCO_3$ | 150–350 | 150–350 |
| Biochemical oxygen demand | mg/L | <5 | <5 |
| Total organic carbon | mg/L as C | 5–25 | 5–25 |
| Total Kjeldahl nitrogen | mg/L as N | <3[c] | 5–30 |
| Total phosphorus | mg/L as P | <1[c] | 0.1–8 |
| Iron | mg/L | 0–0.2 | 0–0.2 |
| Total suspended solids | mg/L | <1 | Below detection limits |
| Turbidity | NTU | <0.2 | <0.1 |
| Silt density index | — | <3 | <3 |
| Bacteria | — | Up to 6 log removal | Up to 6 log removal |
| Viruses | — | Up to 3 log removal | Up to 3 log removal |

[a] CAS = conventional activated sludge.
[b] mg/L = ppm.
[c] Includes preanoxic, postanoxic for nitrogen removal, and chemical addition for phosphorus removal.

processes detailed in Chapters 3 and 4 of this publication. Table 5.2 presents expected product water quality from RO treatment facilities used for advanced posttreatment of tertiary wastewater effluent.

Microfiltration and UF membranes typically reject turbidity, total suspended solids, bacteria, and viruses, as indicated in Table 5.1. Reverse-osmosis membranes have the ability to remove both soluble organic and inorganic constituents. Absent from both tables are EPOCs. Currently, there is little numerical data available because of limited detection methods. However, RO has shown, in several studies, high removal rates (>90%) of EPOCs such as pharmaceuticals, hormones, and industrial chemicals (Kimura et al., 2003), thereby increasing the consideration of this technology for indirect potable water reuse applications or service water discharge applications.

## TREATMENT MECHANISM

Reverse osmosis is a pressure-driven and diffusion controlled membrane process. Osmosis is defined as the passage of a liquid from a weak to a more concentrated solution across a semipermeable membrane. Reverse osmosis is achieved by providing adequate pressure to overcome osmotic pressure so that feed water flows from the more concentrated solution

Table 5.2 Typical Product Water Quality Ranges for RO Treatment Facilities.

| PARAMETER | UNITS | RANGE OF VALUES: SINGLE-PASS | RANGE OF VALUES: TWO-PASS[a] |
|---|---|---|---|
| Total dissolved solids | mg/L[b] | 10–60 | 1–10 |
| Bicarbonate | mg/L as $CaCO_3$ | 10–30 | 2–20 |
| Total organic carbon | mg/L as C | 0.2–1 | 0.1–0.5 |
| Iron | mg/L | 0–0.06[c] | 0–0.04[c] |
| Total Kjeldahl nitrogen | mg/L | ND | ND |
| Total suspended solids | mg/L | ND | ND |
| Turbidity | NTU | ND | ND |

[a] Two-pass refers to a second pass of membrane treatment for the permeate produced in the first pass.
[b] mg/L = ppm.
[c] Additional pretreatment for iron may be required, depending on the form of iron present.

to the "clean" water side of the membrane. The product water (permeate) is collected in tubes and transported for use as high-quality product water. In a concentrate staged system, the concentrate (retentate) stream is passed to subsequent membrane trains for further treatment and disposal. Membrane equipment configurations associated with wastewater processes are discussed in the following subsections.

## EQUIPMENT MANUFACTURERS

There are several membrane system suppliers that are capable of constructing complete membrane treatment systems. Membrane treatment systems to date have ranged from 2.6 to 260 $m^3$/min (1 to 100 mgd) and have varied in configuration based on capacity and manufacturer.

## SAFETY

Operator safety should always be a major priority for the utility. The operator is advised to thoroughly read and review all material provided in operations and maintenance manuals and to follow safety instructions provided by the membrane equipment manufacturer. It is beyond the scope of this publication to address safety aspects related to all available membrane equipment and processes.

Particular care should be taken with the handling of pretreatment and cleaning chemicals and with the operation and maintenance of the high-pressure feed pumps and the pressurized membrane process skids, as pressures typically exceed 1380 $kN/m^2$ (200 psi). Operators should also thoroughly read and review material safety and data sheets provided for each chemical.

# Process Configurations

A standard RO system consists of the membrane process itself and required pretreatment and posttreatment systems. Successful operation of these process systems is highly dependent on feed water quality.

## PRETREATMENT

The feed water supply source and quality will have the largest effect on the life of membrane elements and their ability to be cleaned. Membrane elements, over time, will become either scaled or fouled based on the constituents in the feed water supply source. During design of a membrane treatment facility, a complete and accurate source water quality analysis should be performed to determine adequate chemical and physical pretreatment and the scaling potential of the feed water source. Membrane scaling occurs as the feed water to the membranes becomes more concentrated and various salts tend to precipitate onto the membrane surface. Fouling potential can be determined based on pilot testing or full-scale operation. In the absence of this data, fouling potential may be estimated based on silt density index (SDI), turbidity, biological activity, total organic carbon (TOC), iron, and hydrogen sulfide measurements. For example, RO manufacturers typically require a source water turbidity less than 1 NTU and a SDI <5. A feed water SDI of <3 is preferred.

A membrane treatment facility can be designed, constructed, and operated using the best practices and still have operational difficulties if pretreatment chemicals are not chosen and applied correctly for the conditions of service. The use of the wrong chemical can result in higher feed pressures, more frequent cleaning, lower effluent water quality, reduced membrane life, and ultimately higher operations and maintenance costs. The proper use of chemical and physical pretreatment can extend the life of the membranes and the period of time between cleanings. Pretreatment steps for systems using RO to posttreat secondary effluent may consist of

- Microfiltration or UF,
- Sodium bisulfite chemical addition,
- Chloramination,
- Sulfuric acid chemical addition,
- Scale/threshold inhibitor chemical addition, and
- Cartridge filtration.

## Particulate Control

Microfiltration and UF have emerged as the most suitable pretreatments for advanced posttreatment of wastewater by RO. The use of MF/UF processes for wastewater treatment is detailed in Chapters 3 and 4 of this publication. It should be noted that, if iron is used to aid MF/UF treatment, consideration should be given to an iron-removal process before NF/RO membrane treatment.

Cartridge filtration can provide additional particulate control. Cartridge filtration processes often provide the last line of defense against suspended solids for many RO membrane processes. Cartridge filtration equipment provides removal of suspended particles and helps facilities meet the recommended feed water turbidity and SDI requirements set forth by membrane manufacturers in their warranties.

Proper replacement of cartridge filters can ensure that a membrane facility is not exposed to the pass-through of suspended particles such as clay, silt, sand, or chemical impurities. If these suspended particles are passed on to membrane elements, they will cause irreversible damage to the membrane surface. It is also very difficult to remove such particles from RO membrane elements with standard cleaning practices.

The pore size of cartridge filters ranges from 1 to 25 $\mu$m ($4 \times 10^{-4}$ to $1 \times 10^{-4}$ in.). The design standard for cartridge filtration is typically 15.1 to 18.9 L/min (4 to 5 gpm) per 250 mm (10 in.) of cartridge element. Cartridge filter elements are commonly available in lengths ranging from 250 to 1000 mm (10 to 40 in.). The pressure drop for a clean filter housing loaded with cartridges is approximately 20 kN/m$^2$ (3 psi). Replacement is typically initiated based on increased differential pressure or as a scheduled maintenance item. Cartridge filter elements should be replaced when the differential pressure across the unit reaches 100 kN/m$^2$ (15 psi). Otherwise, the scheduled maintenance would be at the discretion of maintenance staff.

Chlorination/Dechlorination

The use of chloramines to control biological activity has been successfully applied at several RO facilities for advanced posttreatment of wastewater. Chloramination is a form of disinfection that consists of the combination of chlorine with ammonia to form chloramines. Chloramines are stable biocides and oxidants; however, they are slower to react and provide a weaker "kill" than free chlorine or other oxidants. Chloramines are formed by adding ammonia to water that contains free chlorine. Ammonia may be added as ammonia gas or as ammonium chloride. Chloramines at residual levels of 2 mg/L (2 ppm) have been used successfully as pretreatment to RO processes used for water reclamation (Alexander, 2003). In these applications, chloramines have played an important role in controlling biological fouling in membrane wastewater applications.

When residual free chlorine is present in feed water, dechlorination is required. As a strong oxidizing agent, residual free chlorine in feed water must be dechlorinated before use of RO membranes to prevent damage. Though some RO membranes may initially exhibit tolerance to chlorine, eventual degradation may occur after continuous exposure to 1 mg/L (1 ppm) or higher levels of free chlorine, depending on the pH, temperature, and residual transition metals such as iron in feed water. Under alkaline pH conditions, chlorine degrades the membrane faster than at neutral or acidic pH.

Sodium metabisulfite or sodium bisulfite (SBS) is most commonly used for removal of free chlorine, as shown in the following reaction:

$$Na_2S_2O_5 + H_2O \rightleftharpoons 2NaHSO_3$$

$$NaHSO_3 + HOCl \rightleftharpoons HCl + NaHSO_4$$

Other uses of SBS include pretreatment dosing to control biological fouling. Colloidal fouling has also been reduced by this method. Sodium bisulfite can also be helpful for controlling calcium carbonate scaling by supplying protonium ($H^+$) ions as shown below:

$$2NaHSO_3 + CaCO_3 \rightleftharpoons Na_2SO_3 + Ca^{2+} + HCO_3^- + HSO_3^-$$

Because most NF and RO membranes are not chlorine- or oxidant-tolerant, a dechlorination step is required before membrane treatment with NF or RO membranes. Sodium bisulfite is extremely effective in dechlorinating the feed stream; however, it should be noted that this chemical can also act as a nutrient source, contributing to the biofouling of the membranes. Therefore, SBS can, under certain conditions, have a negative effect on RO operations because of the increased potential for biological fouling. Because results from adding SBS can be unpredictable as a result of other variables, it is highly recommended that pilot testing be conducted with the chemical at a specific facility before full-scale use.

Scale Prevention

The addition of a mineral acid to lower the pH of RO feed water is one of the most effective ways to prevent the formation of calcium carbonate ($CaCO_3$). Because the solubility of calcium carbonate salt depends on the bicarbonate ion and pH, converting the bicarbonate ion to carbon dioxide and lowering the pH will increase its solubility. The addition of a mineral acid can also be a method for controlling biological fouling in a RO system. It should be noted, however, that the converted carbon dioxide is present in feed water as a dissolved gas and, as such, is not rejected by RO membranes. The most common mineral acid used for this purpose is sulfuric acid; however, injection of hydrochloric acid (HCl) or carbonic acid is also a viable method for pH adjustment. The use of HCl for pH adjustment contributes to chloride levels in feed water and should be carefully considered if chlorides are constituents of concern in the product water. Attention should also be paid to ensure that storage and feed facilities are designed to handle the fuming that is characteristic of volatile HCl or increased storage volume and pumping capacities needed to use lower-strength HCl.

Sulfuric acid was among the first pretreatment chemicals used for RO technology and, as such, has a well-documented history of successful application in membrane treatment facilities. While the injection of sulfuric acid does reduce the scaling potential for calcium carbonate, as it is injected, the concentration of sulfate ions is accordingly increased in the feed water. This can further contribute to scale formation from calcium, strontium, or barium sulfate salts and should be accounted for when providing water-quality data to a pretreatment chemical manufacturer for a product recommendation.

Sulfuric acid is classified as a strong acid and, as such, requires that proper safety precautions be taken by operating personnel during the handling of acid and the maintenance of the storage and feed facilities. Currently 93 to 98% concentrations of sulfuric acid are commercially available; however, the trend seems to be moving towards standardization at 98%.

Scale inhibitors (antiscalants) are required when the concentration of a salt exceeds its solubility. Precipitation of salts may begin to occur with the formation of crystal, which acts as a catalyst to the formation of further crystal on the surface until the crystals reach a size and density such that they fall out of suspension. This process will continue until the ions left in solution are at their solubility limit. Scale inhibitors are an effective means of preventing the fouling of RO membrane elements as a result of the formation of scale.

These chemicals slow the precipitation process by inhibiting salt growth by their adsorption onto the surface of the forming salt crystal, which, in turn, slows its expansion and prevents the attraction of more supersaturated salt to the crystal structures. Furthermore, some scale inhibitors have some dispersive qualities, which involve surrounding particles of suspended salt or organic solids with the anionically charged scale inhibitor. These anionically charged particles will repel each other to prevent the agglomeration of the particles to larger particles that may precipitate (Nemeth, 1997).

Typical scale inhibitors consist of molecules that contain carboxylic acid (–COOH) or phosphate ($PO_4^{-3}$) functional groups. Polyacrylate molecules (molecular weight distribution from 1000 to 5000 Da) contain multiple carboxylic acid functional groups and are commonly used in many inhibitors for reducing carbonate and sulfate formation.

Polyphosphonates

Polyphosphonates have been used for scale control in boilers and other industrial processes and were among the first scale inhibitors used in RO membrane processes. The functional groups most often used in scale inhibitors contain carboxylic acid (–COOH) or phosphate ($PO_4^{-3}$).

Polyphosphonates consist of sodium hexametaphosphate (SHMP), or sodium hex, and potassium pyrophosphonate. Sodium hexametaphosphate is inexpensive when compared with other scale inhibitors and has been commonly used in conjunction with a mineral acid for scale control. While SHMP has a history of successful operation in membrane processes, care must be taken to avoid hydrolysis of SHMP in the dosing feed tank (a fresh solution should be made every 3 days). Hydrolysis can not only decrease the scale-inhibition efficiency but also lead to formation of calcium phosphate scaling, which causes fouling in the membrane system. Additionally, because it is not as effective for scale control as other newer polymeric scale inhibitors, it typically requires a higher dose. For these reasons, the use of SHMP has decreased significantly in recent years.

Organophosphonates are another form of phosphonates that offer an improvement over SHMP. They are more expensive than SHMP but are more resistant to hydrolysis and offer scale inhibition and dispersion abilities similar to those of SHMP.

Polyacrylates

Reverse-osmosis membrane treatment technologies have benefited greatly from the development of polymeric scale inhibitors. Perhaps the most significant among those available are polyacrylates or polyacrylic acids. Polyacrylates can be divided into two categories: lower-molecular-weight and higher-molecular-weight. Lower-molecular-weight polyacrylates typically have a molecular-weight distribution in the range of 1000 to 5000 Da and contain multiple carboxylic acid functional groups. These polyacrylates are most effective at inhibiting carbonate and sulfate formation. Higher-molecular-weight polyacrylates typically have a molecular-weight distribution in the range of 6000 to 25 000 Da and are most effective at dispersion but not as effective at scale inhibition. In general, polyacrylates are more effective than SHMP. However, precipitation reactions may occur with cationic polyelectrolytes or multivalent cations such as aluminum or iron to foul the membrane.

Blend inhibitors are a combination of low- and high-molecular-weight polyacrylates or a blend of low-molecular-weight polyacrylates and organophosphonates for excellent dispersive and inhibitor performance. Blend inhibitors can offer advantages over monochemical inhibitors that, if overinjected, can result in the inhibitor itself falling out of solution with a multivalent cation. Many of the latest developments in scale-inhibitor and dispersant technology are proprietary chemicals that fall into this category. As technology continues to push the limits of today's RO systems, blend inhibitors seem to offer the most potential for future developments (Nemeth, 1997).

# MEMBRANE TREATMENT

The use of RO membrane processes for the treatment of tertiary wastewater effluent is becoming more prevalent for applications that require a high-quality effluent. Reverse osmosis provides removal of nitrogen, nitrates, heavy metals, TOC, disinfection byproducts, and pathogens, which are all heavily regulated for wastewater reclamation. More recently, it has been discovered that RO processes provide significant removal of EPOCs such as pharmaceuticals and hormones. The removal of these EPOCs is of particular concern for reclaimed municipal water applications such as irrigation, aquifer and surface water recharge, and indirect potable reuse.

## CHEMICAL CLEANING SYSTEMS

Although proper pretreatment facilities are often in place, eventually membrane elements will require cleaning. Cleaning systems are fairly universal in design and construction because they provide the same function regardless of the facility. Cleaning systems are typically designed to clean one stage or train at a time so that foulants are not passed from one stage or train to the next. Additional information on cleaning systems is included in Chapter 2. Figure 5.1 illustrates a typical RO plant cleaning system.

## POSTTREATMENT FACILITIES

Membrane processes typically require a series of posttreatment processes to stabilize the product water. As discussed, pretreatment steps typically

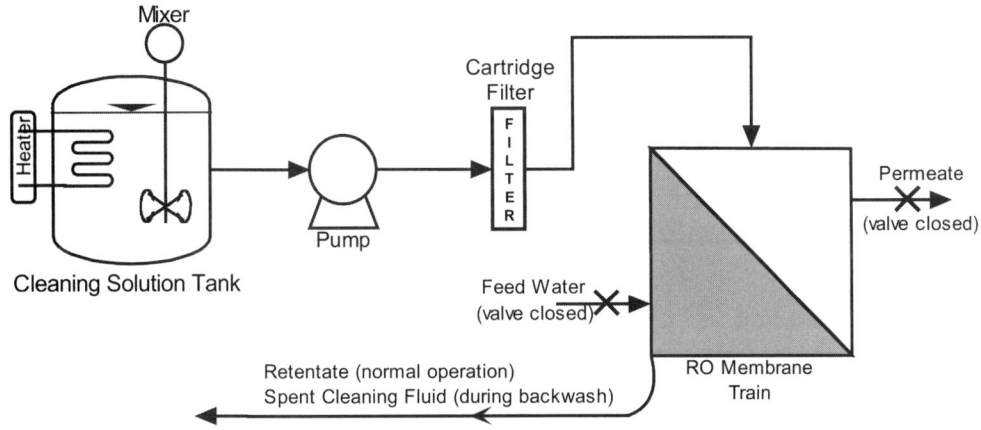

**FIGURE 5.1—REVERSE-OSMOSIS MEMBRANE CLEANING SCHEMATIC.**

include the addition of acids to control scaling. Reverse-osmosis and NF membranes will also remove alkalinity and calcium from feed water. Therefore, product water will often have a low pH and be corrosive. Post-treatment facilities used at RO membrane treatment facilities for treatment of secondary wastewater effluent often include

- Aeration/degasification,
- Stabilization,
- Alkalinity recovery, and
- Disinfection.

Degasification is typically used for the removal of gases from the permeate stream, as membrane processes do not provide removal of gases. Other benefits of this process are an increase in pH and the addition of dissolved oxygen to the product water.

Stabilization of the product water can be provided by the addition of chemicals to the permeate stream. Stabilization chemicals consist of sodium hydroxide (caustic) or hydrated lime to raise the pH. Hydrated lime is a dry powder that is commercially available as a fine powder or as slurry.

Alkalinity recovery can be achieved through blending with other water that is higher in alkalinity. If degasification is used, caustic may be added before degasification to raise the pH and, therefore, convert carbonic acid to bicarbonate. However, increasing the pH is counterproductive relative to the successful removal of some gases; therefore, this treatment step should be considered on a case-by-case basis.

Disinfection can be achieved through the use of gas or liquid chlorine or by several alternative disinfection methods such as addition of chloramines, ozone, and UV disinfection.

## CONCENTRATE DISPOSAL

Disposal of concentrate from NF/RO facilities may present challenges because these systems remove even dissolved contaminants. Depending on the source of the wastewater treated, concentrate from the NF/RO system may have high concentrations of salinity, heavy metals, or other contaminants that can be toxic. Membrane concentrate disposal is highly regulated at the state and federal level. Careful evaluation of the applicable regulations and the cost and sustainability of various disposal options must be made.

Concentrate disposal can be accomplished by surface water discharge, deep well injection, reuse, or concentration methods such as evaporation. Two common municipal practices are surface water discharge and deep well injection. Considerations for the evaluation of concentrate disposal alternatives include

- Cost,
- Regulatory acceptance,
- Concentrate quality, and
- Permitting and monitoring requirements.

The above considerations are very site-specific and, therefore, should be evaluated on a case-by-case basis. The quality and quantity of concentrate can be determined easily once the feed water quality of the source is established. The quantity is easily calculated based on the system recovery rate, and the concentrate characteristics can be approximated based on the feed water quality and the expected concentration factor.

Concentrate disposal from MF/UF pretreatment facilities, if used, also must be managed properly. For more information on the characteristics and the disposal options for MF/UF membrane concentrate, refer back to Chapters 3 and 4 of this publication. Other waste streams such as backwash water from MF/UF systems and spent cleaning solution from RO systems may be returned to the headworks of the upstream wastewater biological process.

# Equipment Configurations

Membrane equipment for RO/NF applications is typically provided in multiple process skids. The design of each process skid is based on establishing a membrane array that is appropriate for the given feed water quality and required treatment capacity. The array is sized based on design criteria for the maximum average flux and maximum lead element flux. These values are typically based on manufacturers' requirements and published design criteria for the membrane industry. Typical flux values for treatment of secondary effluent with RO/NF membranes are between 14 and 20 L/m²·h (8 to 12 gpd/sq ft).

## MEMBRANE PROCESS EQUIPMENT COMPONENTS

Two- and Three-Stage Membrane Array (Concentrate Staged)

Membrane treatment processes are typically provided as staged systems. The array refers to the ratio of pressure vessels and membrane elements per stage. Membrane elements are arranged in stages, with a decreasing number of elements in each subsequent stage. The feed water is sent to the first stage, and permeate is taken from the permeate tube of each vessel as product water. The concentrate from the first stage is used to feed the second stage (likewise from the second stage to the third stage) to recover additional permeate from the concentrate or waste stream. Membrane system recovery for a two-stage system is between 75 and 85%. A three-stage system can achieve between 80 and 92% recovery. Figure 5.2 represents a concentrate staged array.

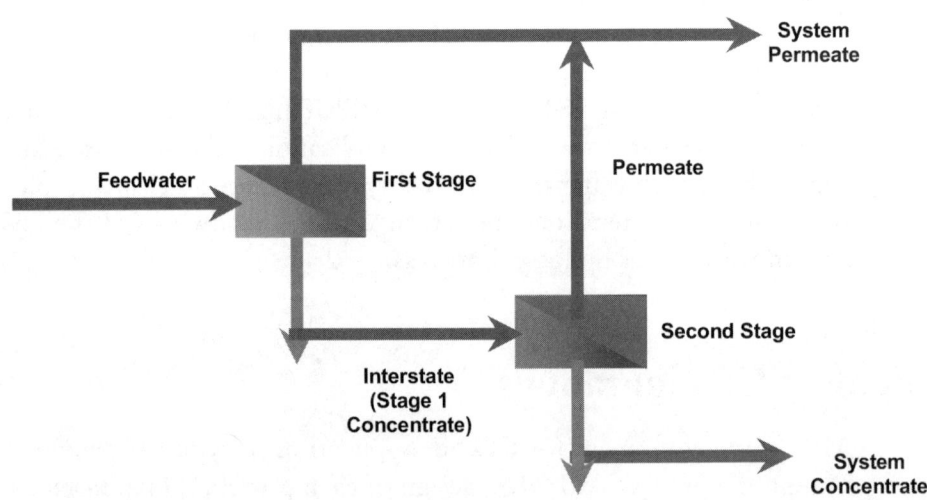

FIGURE 5.2—CONCENTRATE STAGED ARRAY.

### Two-Pass or Partial Two-Pass Membrane Array (Permeate Staged)

A two-pass system is a variation of the two-stage system in which the permeate stream is conveyed to the second pass to provide additional treatment of the product water. This is used when the water-quality requirements of the user or for reuse or disposal are greater than what can be achieved in the first pass alone. Therefore, this configuration is common for ultrapure water applications and indirect potable reuse systems. Depending on finished product water requirements, all or part of the first-pass permeate volume may be passed to the second pass for additional treatment, thereby defining a "two-pass" and a "partial two-pass" system. Figure 5.3 illustrates this two-pass permeate staged concept.

### Membrane Elements

Reverse-osmosis membrane elements are manufactured from two primary materials, cellulose acetate and thin-film composites (TFC) or polyamides. The major element configurations for RO membranes are hollow-fiber and spiral-wound. More details on these materials and element configurations are provided in Chapter 2.

Spiral-wound membrane elements have emerged as the industry standard for RO membrane processes in the United States. The number of membrane elements and the maximum design flux per element define the amount of water that will be produced by a RO membrane system. The

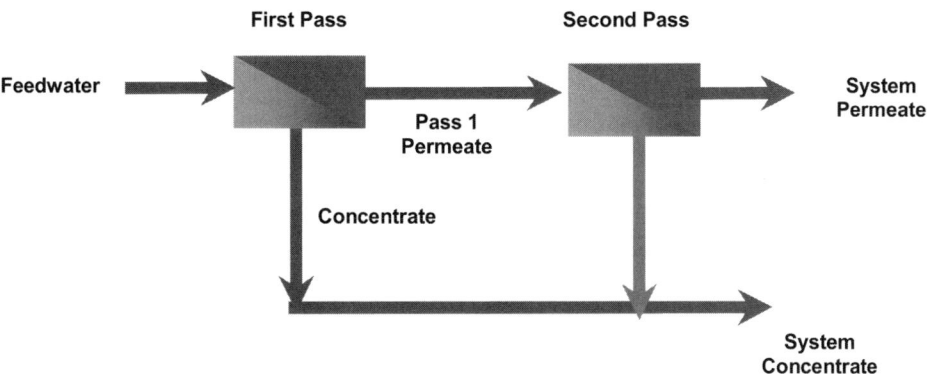

FIGURE 5.3—PERMEATE STAGED ARRAY.

maximum flux is defined by the flux that does not result in excessive system fouling. Recommended flux rates range from 14 to 20 L/m$^2$·h (8 to 12 gpd/sq ft) for treatment of tertiary wastewater treatment effluent.

## TYPICAL INSTRUMENTATION AND CONTROLS

The operation of RO membrane facilities consists of long-term monitoring of performance data rather than short-term process adjustments. Typically, operational data are collected continuously through the facility supervisory control and data acquisition (SCADA) system and collected manually during each shift by onsite operations staff. In the following section, there is a discussion regarding automated and event-driven systems. Automated systems have sufficient monitoring instrumentation and adequate preprogrammed control logic with feedback so that the system can maintain proper setpoints for permeate flow. Event-driven control scenarios for RO systems include startup and shutdown procedures as well as scheduled chemical cleanings. Careful system monitoring will allow for properly scheduled maintenance activities.

Proper chemical feed and recovery setpoints must be maintained to provide sufficient permeate flow. These setpoints are used to provide flux management and fouling control. One of the main causes of membrane system performance degradation or failure is membrane fouling. Therefore, membrane systems should have closely monitored differential pressure increases and/or reduction in permeate flow.

# Operation

This section discusses operational considerations for the RO membrane process only. Typical operational procedures required for standard equipment such as pumps, motors, valves, and other equipment that is common to other treatment facilities should also be considered. This section discusses the effects of feed water chemistry, recovery, and temperature on membrane facility operations. Although proper pretreatment facilities are often in place, eventually the membrane elements will require cleaning. The replacement of membrane elements is a significant operations and maintenance expense; therefore, any action that can be taken to increase the life of the membrane elements should be evaluated and implemented. The provision of adequate cleanings at proper intervals will prolong the life of membrane elements.

## PRETREATMENT CHEMISTRY

The successful operation of a RO membrane process is directly related to feed water quality and the water chemistry associated with the pretreatment regimes that are used. Solubility and the chemistry of slightly soluble salts (also referred to as sparingly soluble salts) are the driving force for most pretreatment chemical reactions in RO membrane processes. The most frequently referenced salt that falls into this category is $CaCO_3$. Calcium carbonate can best be kept in a soluble state through reduction of the feed water pH, which decreases the concentration of bicarbonate available for formation with calcium ions (Kinslow and Manning-Hudkins, 2004).

The most common method for determining the solubility of calcium carbonate in water is the Langelier saturation index (LSI). Waters that have a negative LSI are undersaturated with respect to $CaCO_3$ (tend to dissolve $CaCO_3$), while waters that have a positive LSI are oversaturated with $CaCO_3$ (tend to precipitate $CaCO_3$). The LSI equation is included in Appendix I. Another equation used for determining the solubility of $CaCO_3$ is the Stiff and Davis scaling index (SDSI). The SDSI equation is also included in Appendix I.

## FOULING

The phenomenon of fouling occurs as the result of the accumulation of scale or foulants on the surface of the RO membrane, resulting in impaired performance. However, in practice the term is commonly used to describe either biological or colloidal fouling, whereas mineral scaling (also technically a form of fouling) is commonly referred to as scaling.

Mineral scaling is the result of the precipitation of a slightly soluble salt that occurs when the solubility of that salt is exceeded. Colloidal or particulate fouling can often be differentiated from biological fouling because of the loss of performance in the lead end of a membrane system. Biological fouling can often cause a loss of performance throughout a membrane system but may be most noticeable in the lead elements. Scaling can often be differentiated from other forms of fouling through a loss of performance in the tail end of a membrane system.

### Sulfate Salts

The solubility product ($K_{sp}$) is used to determine the solubility of calcium sulfate ($CaSO_4$) in feed water. If $CaSO_4$ or another sulfate is the limiting salt of concern, the use of a sulfuric acid for pH adjustment will often increase the potential for scale formation by increasing the sulfate ($SO_4^-$) concentration in the water. Therefore, typical means for preventing calcium and/or other forms of sulfate scale is to use a scale inhibitor. The $K_{sp}$ for any given salt is influenced by the feed water temperature and the ionic strength of that salt. The formula for determining the solubility product of a given salt can be found in Appendix I.

Barium sulfate ($BaSO_4$) and strontium sulfate ($SrSO_4$) are other sparingly soluble salts that are formed with sulfates. Similar to $CaSO_4$, the solubility product ($K_{sp}$) is used to determine the solubility of these salts in the concentrate. Most RO membrane manufacturers have projection software that will estimate solubility levels for slightly soluble salts in either a decimal or percentage form so that, for example, a concentrate containing $SrSO_4$ at concentrations two times above its solubility product would be expressed as 200% $K_{sp}$ or 2* $K_{sp}$.

### Silica

Silica fouling can be more difficult to predict and control than other types of fouling. Silica ($SiO_2$) is expressed in terms of milligrams per liter concentration rather than the solubility product method used for sparingly soluble salts. As soluble silica is concentrated to insoluble levels in the RO process, insoluble colloidal silica or silica gel can be formed through a process of polymerization. Methods used to prevent silica from reaching insoluble concentrations include limiting the RO system recovery and increasing the feed water temperature, both of which are difficult and costly to achieve in municipal treatment facilities. While silica solubility is increased at high feed water pH values, this practice can be contradictory to the control of CaCO$_3$ scaling. Recently, the development of new antifoulants has created another alternative that allows silica polymerization to be inhibited while also dispersing particulate matter.

Organics

Organics present in secondary or tertiary effluents may also be present in the original groundwater source (natural organic matter such as humic or fulvic acids) and eventually find their way to the wastewater treatment facility. They may also be soluble microbial products and extracellular polymeric material formed during the biological wastewater process, sometimes referred to as effluent organic matter. Effluent organic matter has been shown to adsorb onto the membrane element surface, causing pore clogging and even a change in membrane surface charge (Drewes et al., 2004).

## INSTRUMENTATION AND CONTROLS

Automated control systems related to RO membrane processes are designed to manage permeate flow, high- and low-pressure events, and high- and low-pH events. Monitoring these process parameters allows proper operation of the essential membrane equipment. These parameters are also monitored so that the warranty of the membrane system or elements is not voided by improper operational practices.

Recovery

Typically, RO membrane processes are operated based on maintaining a percent recovery setpoint. This is accomplished by measuring permeate and concentrate flows and adjusting feed pressure to maintain a constant recovery. Feed pressure may be adjusted using a feed water control valve or the use of variable-frequency drives on pump motors. The recovery of a membrane process has a direct effect on the tendency for slightly soluble salts to become insoluble and begin scale formation. While limiting the recovery can be used to ensure that solubility limits are not exceeded, this practice can result in the waste of a significant portion of the feed water resource. For this reason, dosage of a scale inhibitor (threshold inhibitor) and/or a mineral acid to lower the feed water pH are the most commonly used methods to prevent the formation of scale on the membranes.

Membrane flux is also related to permeate flow as defined earlier in this publication. Typical operating fluxes with NF/RO for advanced posttreatment of wastewater effluent range from 14 to 20 L/m²·h (8 to 12 gpd/sq ft). This is less than the typical operating flux range for water treatment of 20 to 25 L/m²·h (12 to 15 gpd/sq ft). While values within this range are common, the flux for a given system depends on many factors. Significant parameters that influence the flux include temperature, fouling characteristics of the feed water, and pretreatment strategies applied upstream.

### Feed and Differential Pressure

An increase in feed pressure or system differential pressure indicates fouling and scaling because the system requires additional pressure to overcome the resistance to flow. This additional resistance can be remedied by performing a system cleaning. At a minimum, high feed pressures or high system differential pressures should result in an alarm condition and, at the extreme, in an automatic plant shutdown.

### High and Low pH

An unusually high or low pH indicates a pretreatment process upset and should result in an alarm condition. Extreme pH levels outside the warranty conditions should result in an automatic process shutdown.

Table 5.3 presents a typical list of alarms for a RO membrane treatment plant.

## EVENT-DRIVEN SYSTEMS

Event-driven systems related to RO membrane processes include operator-initiated cleaning events, seasonal changes in operation because of flow and load variations, and operational procedures during startup and shutdown events. Membrane performance is often evaluated by two key parameters: mass-transfer coefficient (MTC) and contaminant removal. The MTC accounts for permeate flow, membrane surface area, and energy

Table 5.3 Typical List of Alarms for a RO System (AWWA, 1999).

| Stage | Parameter or Device | Condition | Shutdown | Advisory |
|---|---|---|---|---|
| Pretreatment | Feedwater turbidity | High | X | |
| | Feedwater pH | High, low, deviation | X | |
| | Cartridge filter ΔP | High | | X |
| | Pretreatment chemical day tank level | Low | | X |
| | Pretreatment chemical pump | Failure | X | |
| RO train | High-pressure pump suction pressure | Low | | X |
| | RO bank ΔP | High | X | |
| | Permeate water flow | Low | | X |
| | Permeate conductivity | High | X | |
| | Concentrate flow | Low | X | |
| | Recovery | High | X | |
| | Inlet valve | Close | X | |
| | High-pressure pump | Failure | X | |
| Posttreatment | Permeate pH | High | X | |
| | Permeate pH | Low | | X |
| | Posttreatment chemical day tank level | Low | | X |
| | Posttreatment chemical pump | Failure | | X |

(pressure) requirements and may also be referred to as specific flux. It is an excellent parameter by which to measure the performance and efficiency of the membrane system. It is measured in units of L/m²·h/kPa (gpd/sq ft/psi).

Cleaning Events

The MTC plotted against cumulative operating time provides valuable insight to the performance of the membrane system as a part or a whole. Chemical cleaning frequency is often based on an acceptable level of performance deterioration. A 10 to 15% decline in MTC is often used to indicate the necessity for a chemical cleaning. Typically, the operator initiates a chemical cleaning when treatment capacity requirements are relatively low and the equipment can be safely taken out of operation for the required maintenance or operational cleaning procedure. The MTC will decline with time because, gradually, some degree of irreversible fouling will occur. Once the level of irreversible fouling becomes unacceptable, the membranes must be replaced.

Source Water Changes

Seasonal changes in operation can be expected at biological treatment plants because of influent temperature variations, influent pH values, high flows during springtime snowmelt conditions, high flows during peak rainfall events, or peak loads as a result of seasonal food processing discharge or similar industrial events. Such fluctuations in feed water quality should be addressed so that appropriate changes in feed pressure or chemical pretreatment can be made to optimize performance.

Biofilm Development

Biological monitoring at the membrane system is imperative, as bacteria and algae grow rapidly. Most TFC membranes do not tolerate oxidants such as chlorine. Biological growth, or biofilm, can be deterred by providing permeate flushes if a membrane unit is shut down. If a unit is shut down for more than 2 days, the addition of a biocide should be considered. Bacteria can be measured as a standard plate count, heterotrophic plate count, and TOC. Biological growth in the membrane system will require a timely and adequate chemical cleaning to restore operations.

# CAPACITY MANAGEMENT

During the design of the membrane treatment facility, process reliability or redundancy is typically addressed. Proper redundancy provides the operator with the ability to take some portion of the process equipment out of service for maintenance. Before performing a scheduled maintenance activity, the operator should assess the level of redundancy provided in the treatment system and determine the best possible time to have a tank, equipment train, or pump out of service. This may not be acceptable or possible during peak flow or load conditions. Care should be taken to ensure that any operator-initiated activities that would remove equipment from service are performed when anticipated flows are within the capacity of the remaining equipment.

# PROCESS OPTIMIZATION

Capital and operations and maintenance costs for RO membrane treatment facilities greatly depend on specific site conditions. Site-specific factors that affect construction costs include plant capacity, ancillary plant components such as an administration building, feed water quality, concentrate disposal methods, and pretreatment and posttreatment requirements.

Significant operations and maintenance costs typically consist of power and chemical use, labor, and membrane replacement. Typical energy users in a membrane plant are pumps; blowers for posttreatment aeration, if required; lighting; and controls. Energy-recovery devices (ERDs) can be used to recover energy from the concentrate stream and apply the energy to the feed stream, thus reducing subsequent energy requirements. Chemical use at a membrane plant depends on pretreatment and posttreatment processes that are required. Labor consists of salaries, benefits, and overtime pay. Labor is typically the largest component of the total operations and maintenance cost. Operations and maintenance costs for membrane replacement are typically based on a service life of 5 years, although membrane elements may last significantly longer.

Membrane treatment has benefitted from several technological advances in membrane element design and construction. These advances have allowed membrane treatment to become more affordable. Technological advances in membrane treatment have focused on reducing the operations and maintenance costs associated with the process. Some of these methods are described below:

- *Low-energy membrane elements:* 30 years ago, membrane elements required a great deal of energy because of the high pressure required to overcome osmotic pressures. A research and development effort was made to create membrane elements with a lower hydraulic resistance; as a result, new lower-pressure membrane elements are now available.

- *Energy-recovery devices:* ERDs may be used to provide additional pressure boost to either the first or second stage of a membrane process system. Energy-recovery devices harness the excess pressure in the concentrate line as it leaves the system. The energy from this pressure can then be transferred to any stage or pass of the system to provide a pressure boost without additional electrical input. This can be accomplished using a pressure-exchange device or a free-running turbine. These devices reduce energy consumption and optimize membrane performance in multistage process trains where the pressure in downstream stages is typically underused. Currently, ERDs are not used at any full-scale water reclamation facilities that use RO. However, if the feed pressures required in a system provide adequate excess concentrate pressure, these devices may someday be viable at water reclamation facilities.

- Chemical pretreatment: the proper use of chemical and physical pretreatment can extend the life of membranes and/or the period of time between membrane cleanings. The use of scale inhibitors and pH adjustment with sulfuric acid minimizes scaling of the tail end of the membrane system. During recent years, the development of new scale inhibitors/dispersants has allowed some membrane facilities to run on scale inhibitors alone.

## Routine Monitoring

Adequate operational monitoring will facilitate timely maintenance activities. Routine data monitoring can be manual or automated and continuous. Operational monitoring is required for both feed water characteristics and membrane process conditions. It is also important to monitor the differential pressure across the pretreatment cartridge filter housings to ensure timely replacement of filter elements and minimize the risk of particulate breakthrough.

### FEED WATER

Feed water monitoring typically consists of monitoring turbidity, conductivity, adjusted pH, scale-inhibitor flow, temperature, and SDI. These parameters should be provided on panel-mounted indicators and continuously recorded through the membrane system's SCADA system. Measuring turbidity is important, as it is typically tied to the membrane warranty. Reverse-osmosis membrane manufacturers typically require a feed water turbidity of less than 1 NTU.

Conductivity of the membrane feed water is typically measured as a surrogate for total dissolved solids. A conductivity meter measures the dissolved, charged solids in the water. Measuring the feed water conductivity provides a benchmark of quality that allows plant operations staff to recognize changes. It also allows subsequent performance calculations related to contaminant rejection.

The adjusted pH is monitored to detect changes in feed water quality and to ensure that adequate pretreatment has occurred. As previously discussed, feed water pH is typically adjusted by adding sulfuric acid to minimize $CaCO_3$ scaling. Because there is no standard measurement for detection of scale-inhibitor concentration in feed water, actual scale-inhibitor flow must be measured to ensure the addition of an adequate volume of scale inhibitor.

Silt density index is also monitored as an indication of the fouling potential of feed water. A SDI measurement is also typically required by membrane element manufacturers to maintain the warranty. A SDI test is typically run manually and requires a SDI apparatus, a 500-mL (30.5-cu in.) graduated cylinder, and a stopwatch. The SDI procedure is detailed in several technical publications. The SDIs for secondary treated wastewater are typically greater than 5. The maximum SDI recommended by manufacturers of RO membrane systems is a SDI of 5. Pretreatment with MF and UF membranes typically results in a measurement of 3 or less. A SDI apparatus is illustrated in Figure 5.4 (AWWA, 1999).

The feed water temperature is monitored to allow for the normalization of data relative to a single temperature condition, typically 25° C (77° F). The temperature of the feed water can also have a significant effect on solubility. As temperature increases, the solubility product increases for a water of the same ionic strength. For this reason, cleaning a scaled membrane system is more effective when performed at a higher temperature. Because it is not practical to control the temperature of the feed water, consideration should be made when selecting a scale inhibitor or designing an acid feed system to ensure that

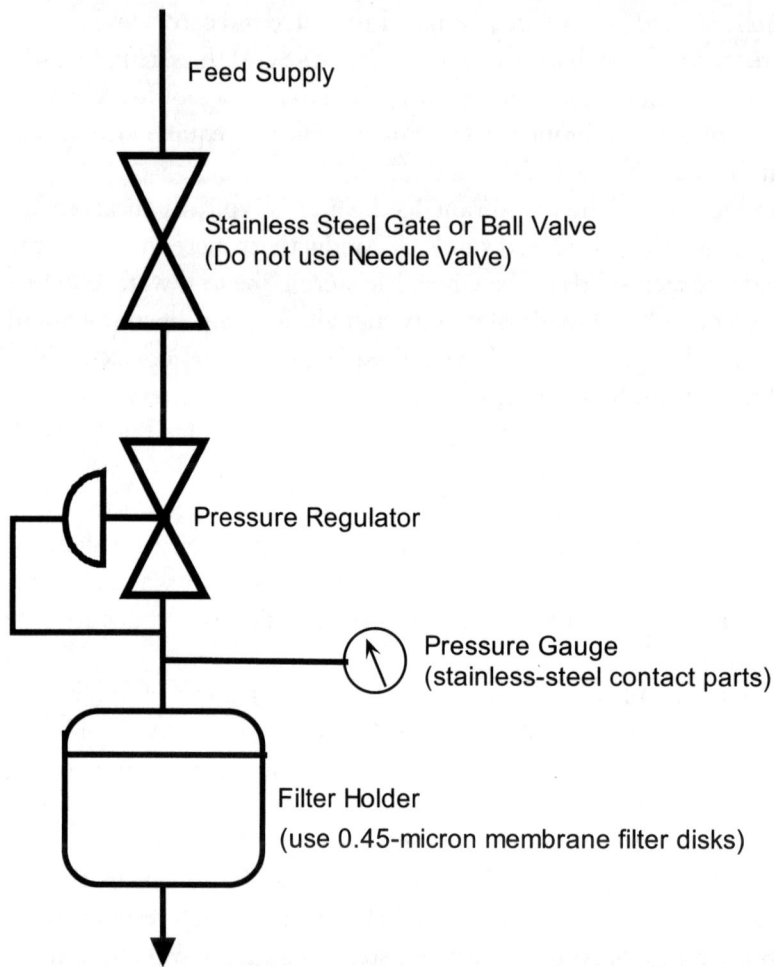

FIGURE 5.4—SILT DENSITY INDEX APPARATUS (micron [μ] = micrometer [μm]) (ADAPTED FROM AWWA, 1999).

slightly soluble salts will not become insoluble at the lower range of feed water temperatures.

## MEMBRANE SYSTEM

Reverse-osmosis membrane system monitoring consists of monitoring permeate and concentrate flow, inlet pressure, permeate and concentrate pH, permeate and concentrate conductivity, and membrane feed pump run time. These parameters are used to assess the overall operational performance of the membrane system. Calculations and data trending related to performance are presented in a following section. The measurement of permeate and concentrate flow in a membrane process train is extremely important, as a membrane system is designed to run at a specified recovery. Producing an excess amount of permeate can result in an elevated factor in the concentrate stream. Measuring flow for two of the three process streams (feed, permeate, and concentrate) can result in the calculation of system recovery.

Pressure should be measured in a minimum of three locations: feed, interstage, and concentrate. This allows for the calculation of pressure drop or differential pressure across the entire system and each stage or pass of the system. System pressure drop is easily calculated as the difference between the feed pressure and the concentrate pressure. An increase in differential pressure over time indicates an increased resistance to flow caused by fouling or scaling. Table 5.4 defines the key parameters that are important to membrane equipment operation.

## DATA INTERPRETATION AND REPORTING

The collection and normalization of membrane process operations data are extremely important for optimizing the performance of the membrane system. Proper monitoring and subsequent maintenance will ensure membrane process longevity. Most membrane treatment facilities collect operational data manually as well as through their SCADA system.

Table 5.4 Reverse Osmosis/NF System Monitoring Requirements.

| PARAMETER | MINIMUM TEST FREQUENCY | RECOMMENDED TEST FREQUENCY | SAMPLE LOCATION |
|---|---|---|---|
| Temperature | 1/day | Continuous | Feed/Perm/Conc |
| pH | 1/day | Continuous | Feed/Perm/Conc |
| Turbidity | 1/day | Continuous | Feed |
| Conductivity | 1/day | Continuous | Feed/Perm/Conc |
| SDI | 1/day | Continuous | Feed |

These data can then be normalized to trend system operations during a variety of seasonal and operational conditions. Normalization programs can be obtained from most membrane manufacturers and incorporated to the membrane treatment plant's SCADA system. Typical parameters for normalization include system flux, feed pressure, and normalized permeate flow. There are several equations that can be used to trend flow or flux of the membrane system over time. Two of these calculations are the calculation of flux and the calculation of recovery. These formulas are included in Appendix I.

The three membrane performance parameters that should be continuously monitored and recorded are percent contaminant rejection, MTC or net specific flux, and differential pressure. When degradation of membrane performance is realized, adequate cleanings and possibly even adjustments in pretreatment strategies should be considered.

### Percent Rejection

Percent rejection for a given membrane element will define the water quality of the permeate stream based on the feed water quality. Reverse-osmosis membrane elements are constructed to provide 90 to 99% rejection of salt, and NF membrane elements are constructed to provide between 75 and 85% of salts. Percent rejection for other contaminants of interest varies, depending on membrane element type and manufacturer. A decrease in rejection would potentially be attributed to scaling, fouling, or mechanical damage to the membrane. The formula for percent rejection and a practical example are included in Appendix I.

### Mass-Transfer Coefficient (Specific Flux)

Mass-transfer coefficient, or specific flux, provides a ratio of flux to TMP. Flux, as discussed previously, represents productivity per membrane surface area. The MTC, therefore, provides a specific functional value, which would illustrate trends in productivity and changes in pressure.

### Differential Pressure

Differential pressure is also a significant operational parameter to monitor. An increase in pressure drop across the RO system or either stage of the system indicates some degree of fouling and scaling because additional pressure is being required to overcome the resistance to flow. This additional resistance can be remedied by performing a system cleaning.

# Maintenance

## CHEMICAL CLEANINGS

Chemical membrane cleaning for RO systems is a bit more complicated than for low-pressure (MF/UF) systems because scaling must be dealt with as well as biological and particulate fouling. Although the life span of RO membrane elements may be difficult to predict, it is possible to prepare for the eventual scaling and/or fouling of the membranes. If cleanings are performed at the correct time and with the appropriate methods, performance can be restored and membrane life can be extended. Utilities using RO processes for treatment of tertiary effluent should consult with other utilities that are similar to their plant before performing their first cleaning. Membrane treatment plant operators possess an expansive knowledge of effective cleaning practices, and it is important that such information is solicited during the design and operation of new membrane plant facilities. Some of the most valuable information gained relative to cleanings will probably result from pilot testing or full-scale operations, where all conditions are specific to the given site.

Reverse-osmosis membrane cleanings are performed at relatively low pressures. The following cleaning agents may be used:

- Acids,
- Bases,
- Detergents or proprietary chemicals, and
- Enzymes.

Chemicals used for cleaning must be compatible with the membranes and construction materials. Cleaning solutions may be heated to maximize their effect; however, the manufacturer's guidelines should be consulted to determine the maximum allowable temperature. Cleaning solution volumes or feed rates are typically estimated at 2.5 L/s (40 gpm) per seven-element membrane pressure vessel. Cartridge filtration should be provided with 5-μm ($2 \times 10^{-4}$-in.) cartridge filters to remove particulate returned in the cleaning system or introduced by cleaning chemicals. Returned cleaning solution should be sampled and analyzed so that additional information relative to scaling and fouling can be gained. Knowing the nature of membrane foulants is necessary to select the proper cleaning chemical. Foulants are typically mixtures of two or more of the following constituents:

- Scale,
- Organisms that make up the biomass,
- Organic debris,
- Colloids,
- Metal hydroxides, and
- Chemical precipitates.

Properly timed and performed cleanings will extend the useful life of the RO membrane elements. Cleaning should be initiated if the differential pressure across either or both stages increases by 15%.

## INTEGRITY TESTING

Integrity testing is used for all types of membrane systems to verify the mechanical integrity of the system. However, because RO membranes remove dissolved solids from the feed water, the permeate from these systems has a very low conductivity. Therefore, if monitoring data show a sudden increase in permeate conductivity, it is possible that a mechanical failure has occurred. The first step in this process is to determine which stage is allowing additional contaminant passage relative to the startup or original profile data. Once the stage has been identified, each vessel in that stage can be sampled to determine which vessels are producing permeate with the highest conductivities. Next, those vessels that have high conductivities should be subjected to probing.

Probing consists of inserting plastic tubing to the product water tube and sampling conductivities at various lengths along the membrane vessel. These length intervals are typically marked on the tubing and represent the location of the various connection points for the membrane elements inside the vessel. As the probe position is moved from the upstream element to the downstream element, the conductivity readings should increase uniformly. Therefore, a sudden or dramatic increase in conductivity will alert the operator of a mechanical failure of the membrane element or membrane or permeate connector O-rings. A typical permeate probing apparatus for a spiral-wound membrane element is illustrated in Figure 5.5.

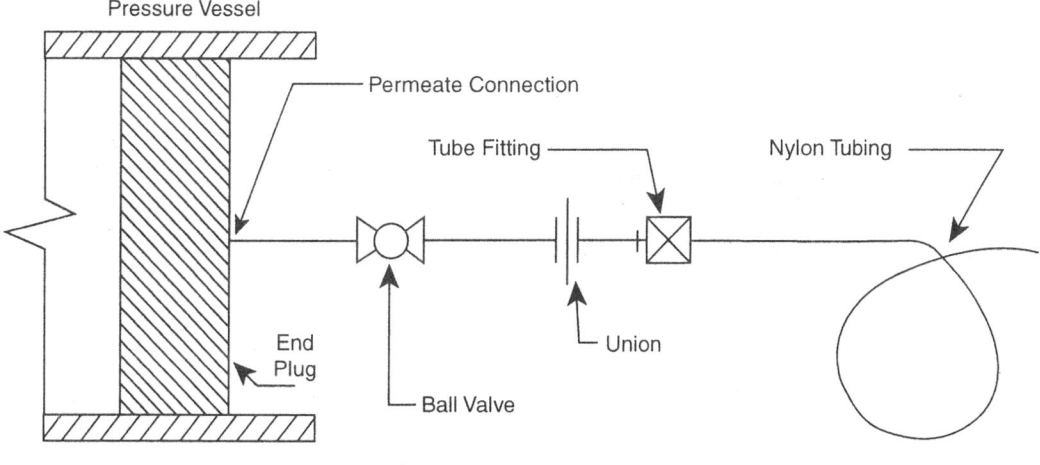

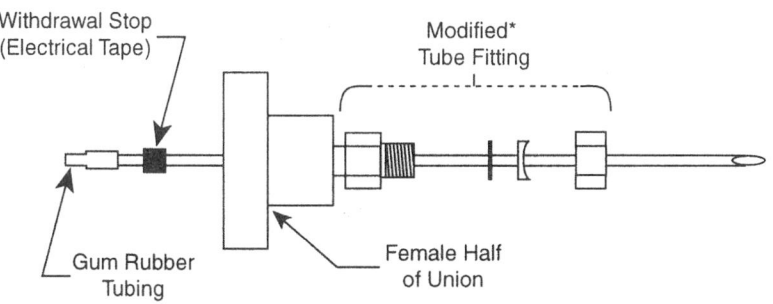

*Tube fitting modified by extending the 1/4-in. bore through the body and discarding the gripper ring.

FIGURE 5.5—PERMEATE PROBING APPARATUS FOR A SPIRAL WOUND MEMBRANE ELEMENT (in. × 25.40 = mm) (AWWA, 1999).

Membrane replacement frequency varies by facility. Membrane elements are typically assigned an average life span of 5 years. There are several facilities that have operated with the same membrane elements for more than 10 years. Eventually, membrane elements are not capable of recovering to an acceptable percentage of their original performance through cleanings and require replacement.

## INSTRUMENT CALIBRATION

As previously mentioned, the operation of RO membrane facilities consists of long-term monitoring of performance data rather than short-term process adjustments required by operations staff. Therefore, the accurate collection of data will lead to a smooth-running facility. These data are typically collected by inline or mounted instruments that must be calibrated periodically for accuracy. The instrument manufacturer and operations staff should develop a routine calibration schedule for the various instruments used at their membrane facility. Instruments reading parameters such as pH and conductivity may be calibrated using hand-held instruments. Plant flows may be calibrated using bucket-testing methods or by evaluating the mass balance of process stream flows for feed water, permeate water, and concentrate water. Calibration methods and schedules should be located in an RO facility's operations and maintenance manual.

# Troubleshooting

Troubleshooting for RO membrane systems can be greatly facilitated by the proper collection of operational data. An increase in first-stage differential pressure typically indicates fouling; an increase in second-stage differential pressure typically indicates scaling of the membrane elements. These and other indicating symptoms are provided in Table 5.5.

Table 5.5  Reverse Osmosis/NF Membrane Process Troubleshooting Guide (Courtesty of the Dow Chemical Company).

| Symptom | Possible Problems |
| --- | --- |
| Sudden ↑ in ΔP, no change in rejection | Clogged cartridge filter, blocked pipes or front end membrane element |
| Sudden ↓ in ΔP, ↓ in rejection | O-ring or brine seal failure, cracked permeate tube |
| Gradual ↓ in normalized permeate flow (NPF) 1st-stage, slight ↑ rejection | Biological or particulate fouling |
| Gradual ↓ NPF 2nd-stage, ↓ 2nd-stage rejection | Scaling |
| Gradual ↑ NPF 1st-stage, ↓ 1st-stage rejection | Lead end degradation caused by reaction of $Cl_2$ with transition metals, or advanced biofouling |
| Gradual ↑ NPF 2nd-stage, ↓ 2nd-stage rejection | Advanced scaling |

## CHAPTER SIX

# Membrane System Case Studies

Introduction ................................................................. 185
Case Study No. 1—Running Springs Water Recycling Plant ................. 186
    Background ............................................................. 186
    Process Overview ....................................................... 188
    Plant Data and Design Summary .......................................... 190
    Highlights ............................................................. 191
Case Study No. 2—The Hamptons Water Recycling Facility ................. 193
    Background ............................................................. 193
    Process Overview ....................................................... 193
    Plant Data and Design Summary .......................................... 196
    Highlights ............................................................. 197
Case Study No. 3—Traverse City Regional Wastewater Treatment Plant ..... 198
    Background ............................................................. 198
    Process Overview ....................................................... 200
    Plant Data and Design Summary .......................................... 202
    Highlights ............................................................. 202
Case Study No. 4—Key Colony Water Reuse Plant .......................... 202
    Background ............................................................. 203
    Process Overview ....................................................... 205
    Plant Data and Design Summary .......................................... 207
    Highlights ............................................................. 207

Case Study No. 5—West Basin Water Recycling Plant . . . . . . . . . . . . . . . . . . . . . . . 208
    Background . . . . . . . . . . . . . . . . . . . . . . . . . . . . . . . . . . . . . . . . . . . . . . . . . . 208
    Process Overview . . . . . . . . . . . . . . . . . . . . . . . . . . . . . . . . . . . . . . . . . . . . . 209
        Microfiltration/Double-Pass Reverse-Osmosis Process Description . . . . . . . 211
        Product Water Quality . . . . . . . . . . . . . . . . . . . . . . . . . . . . . . . . . . . . . . 212
    Plant Data and Design Summary . . . . . . . . . . . . . . . . . . . . . . . . . . . . . . . . . 212
    Highlights . . . . . . . . . . . . . . . . . . . . . . . . . . . . . . . . . . . . . . . . . . . . . . . . . . . 212
Case Study No. 6—Chandler, Arizona, Reverse-Osmosis Facility Wastewater
               Reclamation Plant . . . . . . . . . . . . . . . . . . . . . . . . . . . . . . . . . . . . . 214
    Background . . . . . . . . . . . . . . . . . . . . . . . . . . . . . . . . . . . . . . . . . . . . . . . . . . 214
    Process Overview . . . . . . . . . . . . . . . . . . . . . . . . . . . . . . . . . . . . . . . . . . . . . 216
    Plant Data and Design Summary . . . . . . . . . . . . . . . . . . . . . . . . . . . . . . . . . 218
    Highlights . . . . . . . . . . . . . . . . . . . . . . . . . . . . . . . . . . . . . . . . . . . . . . . . . . . 220
Case Study No. 7—Scottsdale Water Campus . . . . . . . . . . . . . . . . . . . . . . . . . . . . 222
    Background . . . . . . . . . . . . . . . . . . . . . . . . . . . . . . . . . . . . . . . . . . . . . . . . . . 222
    Process Overview . . . . . . . . . . . . . . . . . . . . . . . . . . . . . . . . . . . . . . . . . . . . . 222
        Microfiltration . . . . . . . . . . . . . . . . . . . . . . . . . . . . . . . . . . . . . . . . . . . . 224
        Reverse Osmosis . . . . . . . . . . . . . . . . . . . . . . . . . . . . . . . . . . . . . . . . . . 224
        Posttreatment Facilities . . . . . . . . . . . . . . . . . . . . . . . . . . . . . . . . . . . . .225
    Plant Data . . . . . . . . . . . . . . . . . . . . . . . . . . . . . . . . . . . . . . . . . . . . . . . . . .226
    Highlights . . . . . . . . . . . . . . . . . . . . . . . . . . . . . . . . . . . . . . . . . . . . . . . . . . . 227

# Introduction

In simple terms, membranes used in wastewater treatment fall into one of three categories: nanofiltration (NF)/reverse osmosis (RO), microfiltration (MF)/ultrafiltration (UF), or membrane bioreactor (MBR). In addition, combinations of these technologies are often used to meet the demands of a given project. To illustrate how and where membranes are currently used, the following case studies have been compiled to offer examples of the evolving membrane market in wastewater treatment.

To navigate the case studies, use Figure 6.1. Some of the salient points of interest for each plant are listed as a quick reference.

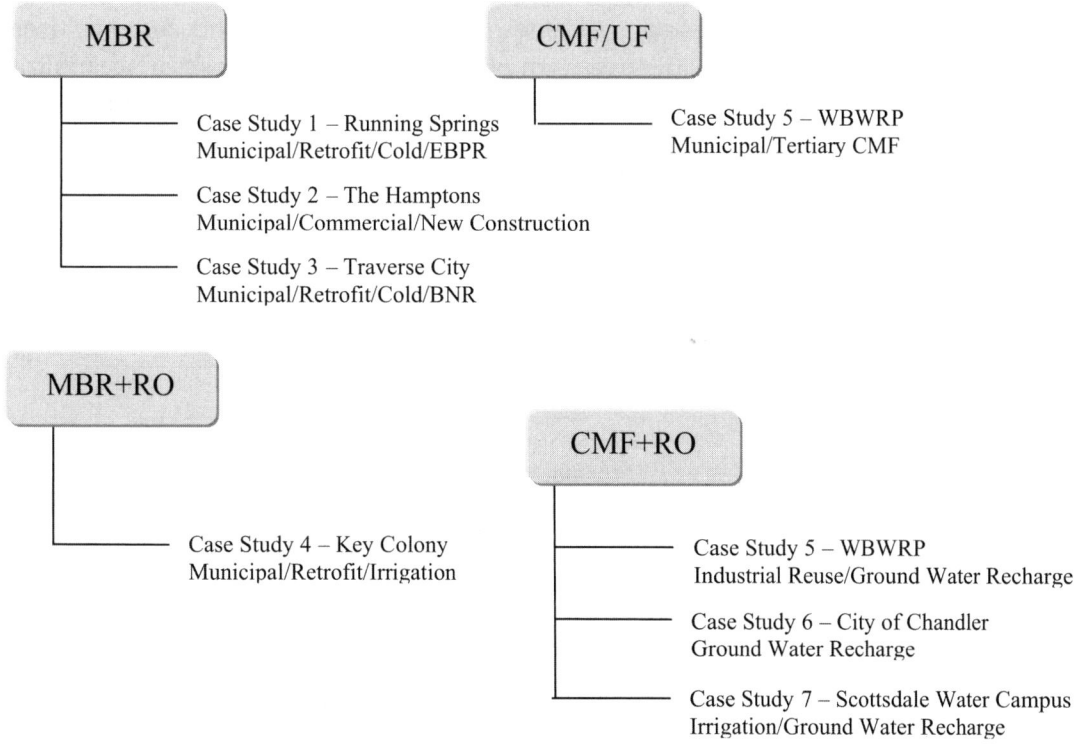

FIGURE 6.1—CASE STUDY NAVIGATIONAL CHART.

# Case Study No. 1—Running Springs Water Recycling Plant

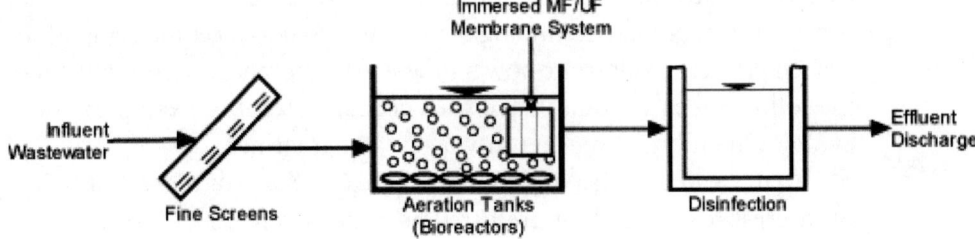

## BACKGROUND

The Running Springs, California, County Water District currently operates a wastewater treatment plant that discharges to subsurface infiltration ponds and a small stream, both of which fall under the jurisdiction of the U.S. Forest Service (see Table 6.1). In 2002, the District was informed that its permit would soon expire and that the new permit would include strict limits on nitrogen and phosphorus.

Table 6.1 Running Springs Water Recycling Plant.

| POINTS OF INTEREST | DESCRIPTION | NOTES |
|---|---|---|
| Location | Running Springs, California | Located in Big Bear National Park |
| Year of system startup | 2003 | |
| Process configuration | MBR | Enviroquip UNR™ process |
| Membrane equipment configuration | Double-deck flat plate | Kubota submerged membrane units |
| Permit type | Reuse | Issued by Department of Forestry |
| Design flow/peak factor | 0.6 mgd/2* | 24-h peak factor |
| Unique features | Gravity operation SymBio® retrofit | Variation on the UNR™ process |

* 0.6 mgd/2 = 1136 m³/d.

To address the impending nutrient limits, the district intially tried increasing the plant sludge age by increasing the operating mixed liquor suspended solids (MLSS) concentration in the aeration tanks. However, existing rectangular secondary clarifiers could not handle the increased loading, and the District had to explore other options.

Starting in April 2003, the plant was retrofitted to a MBR system, including a process to promote enhanced biological phosphorus removal (EBPR). The upgrade of the plant was carried out in stages by converting exisiting clarifiers and aeration basins (see Figure 6.2). By September 2003, the plant was completely converted and optimized to meet future nutrient limits.

FIGURE 6.2—RUNNING SPRINGS RETROFIT TO A MBR.

## PROCESS OVERVIEW

The Running Springs Water Recycling Facility (WRF) uses a unique treatment strategy called the UNR™ process (adapted from Trivedi, 2004). Membranes are used for filtration of particulate material in lieu of clarifiers. The aerobic portion of the plant is operated at low dissolved oxygen (LDO) content using the SymBio® process to promote simultaneous nitrification/denitrification (SNdN). The wastewater is fed first to an anoxic zone before flowing by gravity into an anaerobic zone.

As shown in Figure 6.3, raw wastewater enters the plant through a fine screen to remove particles greater than 0.32 cm (0.12 in.) diameter. The screened influent is then degritted in an aerated grit chamber before being fed to the anoxic (AX) basin.

Mixed liquor overflows from the constant level AX basin into the anaerobic/anoxic basin (AN/AX) for removal of nitrates and promotion of EBPR. A second function of the AN/AX basin is to partially equalize incoming flow and allow for optimization of hydraulic performance. Submersible pumps continuously recycle mixed liquor to the biological

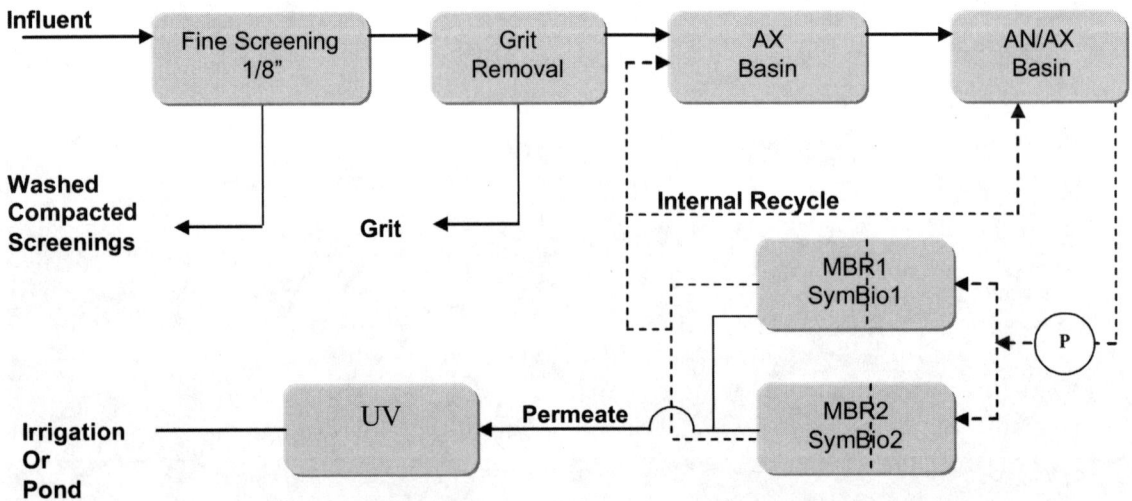

FIGURE 6.3—PROCESS FLOW DIAGRAM FOR THE RUNNING SPRINGS WRF (in. × 25.4 = mm).

reactors at roughly four times design flow or 9464 m$^3$/d (2.5 mgd). The reactors are partitioned into MBRs and LDO zones.

Whatever is not filtered by the membranes returns to the AX basin as thickened sludge to complete an internal recycle loop. To control sludge age, mixed liquor is periodically wasted from the system and dewatered before disposal. Filtrate is returned to the head of the plant for treatment.

Membrane performance is optimized by automatically matching permeate flow to hydraulic demand while minimizing changes in flux. Hydraulic demand is approximated by looking at the level in the AN/AX basin, which is divided into four operator-adjustable flux (flow) bands. These adjustable level bands are assigned flow setpoints that correspond to no-flow, low-flow, design-flow, and peak-flow conditions. Therefore, during any given day as flow to the plant outpaces the rate of treatment, the level in the AN/AX basin increases until the plant is automatically placed into peak-flow mode. As flow to the plant subsides later in the evening, the level in the AN/AX basin will decrease, and eventually the plant will automatically transition to no-flow mode or what is called intermittent mode. Using the level bands smooths out flux setpoint changes and allows membranes to be used according to demand.

Each MBR contains eight double-deck Kubota submerged membrane units and separate fine-bubble diffusers. The fine-bubble aeration system is used to supplement the process oxygen demand and control the dissolved oxygen concentration. The MBR basins are components of the UNR™ process and are often referred to as membrane basins.

Using real-time nictotinamide adenine dinucleotide (NADH) and dissolved oxygen signals, the process controller automatically adjusts the speed of a dedicated blower to keep the dissolved oxygen concentration in the dual-function MBRs between 0.2 and 0.8 mg/L to ensure that SNdN occurs. From April 2003 to August 2003 the plant was operated based on dissolved oxygen measurements and conventional treatment strategy. During September 2003, when the process was optimized, average total nitrogen levels decreased from 14.4 to 3.9 mg/L (14.4 to 3.9 ppm) and total phosphorus levels decreased from 7.4 to 4.9 mg/L (7.4 to 4.9 ppm).

At the Running Springs WRF, the peak factor coupled with limited equalization requires that membranes be able to handle high fluxes,

>50.1 L/m²·h (>29.5 gpd/sq ft), for up to several hours a day and, in some cases, for 24 consecutive hours. Operational data from November 2004 (Table 6.2) demonstrate the ability of membranes to sustain a flux of 30.6 L/m²·h (18 gpd/sq ft) at temperatures ranging between 11.4° and 12.7°C (52.5 and 54.9°F).

The operating flux recommended by the supplier is 22.6 L/m²·h (13.3 gpd/sq ft) at 12° C (53.6°F); however, operation at higher fluxes is permissible, provided that the transmembrane pressure (TMP) does not exceed 20.7 kPa (3 psi). Throughout most of November 2004, the TMP increased by approximately 3.7 kPa (0.53 psi) to a total value of 6.1 kPa (0.89 psi), assuming 2.1 kPa (0.3 psi) of pipe loss.

To handle extended peaks during extreme cold weather conditions, <10°C (50°F), a flux-enhancing polymer is occassionly added to increase the critical flux. Data graphed in Figure 6.4 illustrate that a critical flux of greater than 50.9 L/m²·h (30 gpd/sq ft) was achieved following the addition of 400 mg/L (400 ppm) of polymer during a trial conducted in December 2004.

## PLANT DATA AND DESIGN SUMMARY

Tables 6.3 and 6.4 provide information about the design parameters, results, and requirements at Running Springs.

Table 6.2  Running Springs Hydraulic Performance Data.

| DATE | NET FLUX[a,b,c] (gpd/sq ft) | TMP (psi)[d] | TEMPERATURE (°C)[e] |
|---|---|---|---|
| 11/05/04 | 18.0 | 0.66 | 12.7 |
| 11/07/04 | 18.0 | 0.65 | 12.4 |
| 11/14/04 | 18.0 | 0.82 | 12.0 |
| 11/20/04 | 18.0 | 0.98 | 11.4 |
| 11/24/04 | 18.0 | 1.19 | 12.0 |

[a] Fluxes not corrected for temperature.
[b] Net values based on a 1 min/10 min relax cycle.
[c] 18 gpd/sq ft = 30.6 L/m²·h (1 gpd/sq ft = 1.698 L/m²·h).
[d] Conversion from psi to kPa: multiply psi by 6.89.
[e] Conversion for °C to °F: multiply °C by 1.8 and add 32.

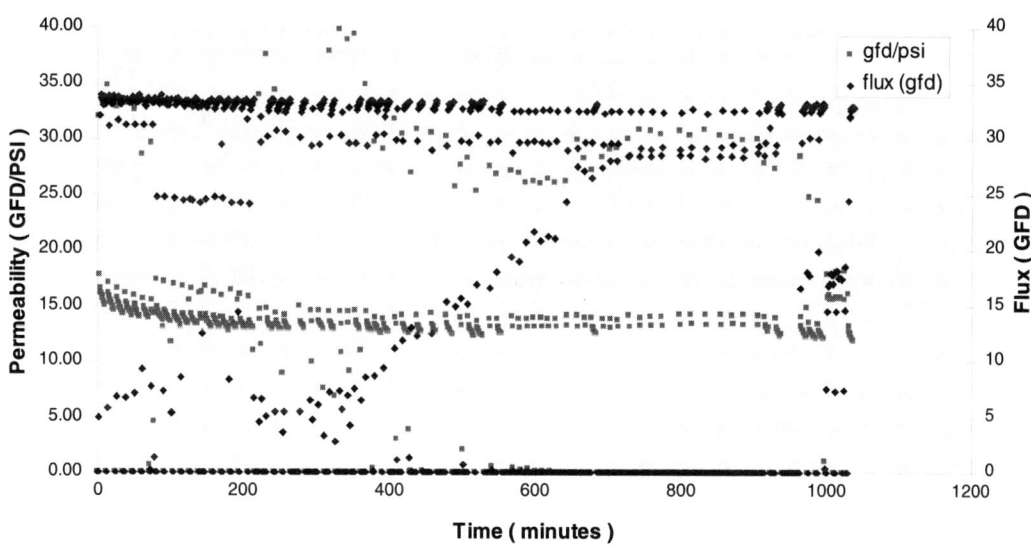

FIGURE 6.4—FLUX AND PERMEABILITY WITH FLUX ENHANCER AT 13.3°C (GFD = gpd/sq ft × 1.698 = L/m²·h; GFD/PSI × 0.246 = $\frac{L/m^2 \cdot h}{kPa}$ ).

## HIGHLIGHTS

We have learned the following from this case study:

1. Monitoring NADH allows for process optimization beyond dissolved oxygen control. Dissolved oxygen probes register changes in water phase concentration but do not reflect biological activity. Following optimization, total nitrogen levels decreased from 14.4 to 3.9 mg/L (14.4 to 3.9 ppm) and total phosphorus levels decreased from 7.4 to 4.9 mg/L (7.4 to 4.9 ppm).

2. Poor screening can degrade membrane performance by reducing air scour efficiency. Instantaneous peak flows must be considered while sizing headworks to avoid carryover or fugitive debris from entering the MBR.

3. Sustainable cold weather—12°C (53.6°F)—fluxes have been recorded at >30.6 L/m²·h (>18 gpd/sq ft). Adding a specially formulated coagulant increased critical flux to 49.9 L/m²·h (29.4 gpd/sq ft) net value to accommodate extended peak-flow conditions.

Table 6.3  Running Springs Membrane Filtration Process Design Parameters.

| Parameters | Units | Current | Design (Future) |
|---|---|---|---|
| Maximum month average daily flow | mgd[a] | 0.60 | 1.0 |
| Peak day flow | mgd | 1.2 | 2.0 |
| Minimum day average process temperature | °F[b] | 50 | 50 |
| Maximum day average process temperature | °F | 71 | 71 |
| Annual average flux | gpd/sq ft[c] | 14.7[d] | 14.7 |
| Peak hour average flux | gpd/sq ft | 29.4 | 29.4 |
| Membrane cassette type | — | EK300 | EK200 |
| Number of cartridges per cassette | — | 150 | 100 |
| Surface area per cassette[e] | sq ft | 1291 | 1291 |
| Number of cassettes installed per train | — | 8 | 16 |
| Number of trains per MBR (membrane tank) | — | 2 | 2 |
| Number of MBRs (membrane tank) | — | 2 | 2 |

[a] To convert from mgd to m³/d, multiply mgd by 3785.
[b] To convert from °F to °C, subtract 32 from °F and multiply by 0.556.
[c] To convert from gpd/sq ft to L/m²·h, multiply gpd/sq ft by 1.698.
[d] Assuming an operating temperature of 15°C (59°F).
[e] To convert from sq ft to m², multiply sq ft by 0.093.

Table 6.4  Running Springs Treatment Results/Requirements.

| Parameter[a] | Average Influent | Average Effluent[b] | Anticipated Future Limits |
|---|---|---|---|
| $CBOD_5$ | 223 mg/L[c] | <2.0 mg/L | <5 mg/L |
| TSS | 165 mg/L | <1.0 mg/L | <5 mg/L |
| TKN | 41 mg/L | <1.0 mg/L | Not permitted |
| Total nitrogen | Not available | <4.0 mg/L | Not permitted |
| TIN | Not available | Not available | <5 mg/L |
| Phosphate | 15 mg/L | <5.0 mg/L | <2 mg/L |
| Turbidity | — | Not available | <0.2 NTU |

[a] $CBOD_5$ = carbonaceous five-day biochemical oxygen demand; TSS = total suspended solids; TKN = total Kjeldahl nitrogen.
[b] Effluent values based on samples taken during September 2003.
[c] 1 mg/L = 1 ppm.

# Case Study No. 2—The Hamptons Water Recycling Facility

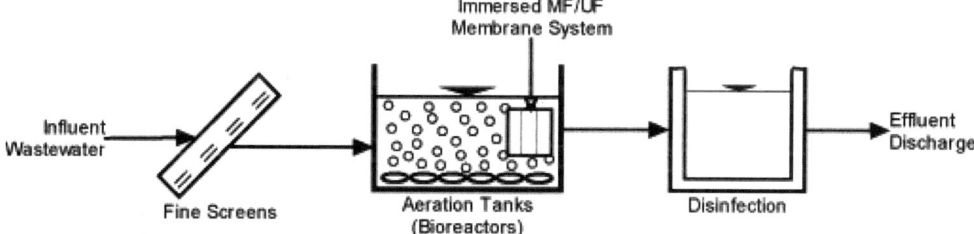

## BACKGROUND

The Hamptons WRF (see Figure 6.5 and Table 6.5) was constructed to provide the necessary utility service to a large planned development and supplement the Forsyth County, Georgia, service area. Within the 2.08-km² (514-ac) development are a championship golf course and a 480-unit subdivision. The effluent from this facility is used for irrigation at the golf course, residential homes, and other green spaces within the development.

The Hamptons WRF primarily treats the wastewater from the on-site subdivision and the golf course clubhouse. However, provisions are in place and capacity is allocated to also provide service to Forsyth County and adjacent developments.

## PROCESS OVERVIEW

From system startup in 2003, influent flow rates have ranged between 189 and 454 m³/d (0.05 and 0.12 mgd). As a result, only one of the four MBR trains is currently online. The plant was designed to allow for expansion by installing additional membrane units as demand increases. The owner expects additional reactors to be brought online within the next year.

Wastewater contaminant concentrations are higher than typical municipal wastewater values as a result of the food service contribution from the clubhouse and minimal infiltration in the new collection system. To account for potential cold weather conditions, additional membrane capacity is provided for in the design. The additional membrane capacity coupled with the deep tanks necessary for gravity operation eliminated the

**FIGURE 6.5—THE HAMPTONS WATER RECLAMATION FACILITY.**

need for dedicated aeration zones and simplified the process flow diagram without reducing treatment efficiency. In fact, data in Figure 6.6 illustrate the ability of the system to reduce effluent nitrogen concentrations (nitrate and total Kjeldahl nitrogen [TKN]) to approximately 5 mg/L (5 ppm) using a single recycle stream and the MBR as the aerobic reactor.

At the present time, the plant is being operated at high MLSS concentrations and long solids retention times (SRTs) to reduce solids handling costs and associated maintenance costs. In general, the MBR MLSS concentration is kept at approximately 25 000 mg/L (25 000 ppm) and the net flux is kept at approximately 17 L/m$^2$·h (10 gpd/sq ft). A graph of typical MLSS values is shown in Figure 6.7. Because this is a gravity plant, the permeability must be maintained at a relatively high level

Table 6.5 The Hamptons Water Recycling Facility.

| POINTS OF INTEREST | DESCRIPTION | NOTES |
| --- | --- | --- |
| Location | The Hamptons, Georgia | Located on a golf course near the club house |
| Year of system startup | 2003 | |
| Process configuration | MBR | Enviroquip process |
| Membrane equipment configuration | Double-deck flat plate | Kubota submerged membrane units |
| Permit type | Reuse discharge | Issued by Georgia Environmental Protection Department |
| Design flow/peak factor | 0.9 mgd/2[a] | 24-hr peak factor |
| Unique features | Gravity operation New construction | Homeowner irrigation in service area |

[a] 0.9 mgd/2 = 1703 m$^3$/d.

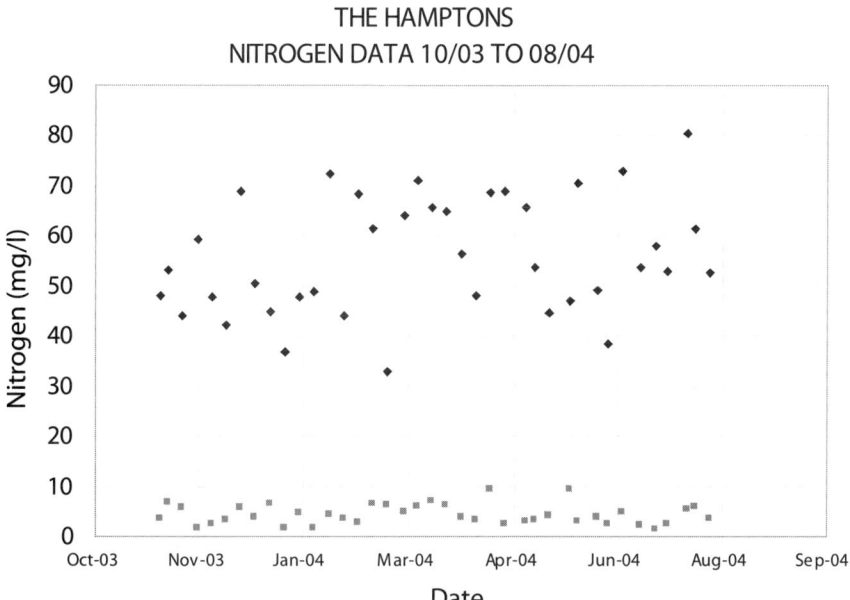

FIGURE 6.6—NITROGEN DATA FROM THE HAMPTONS WRF (mg/L = ppm).

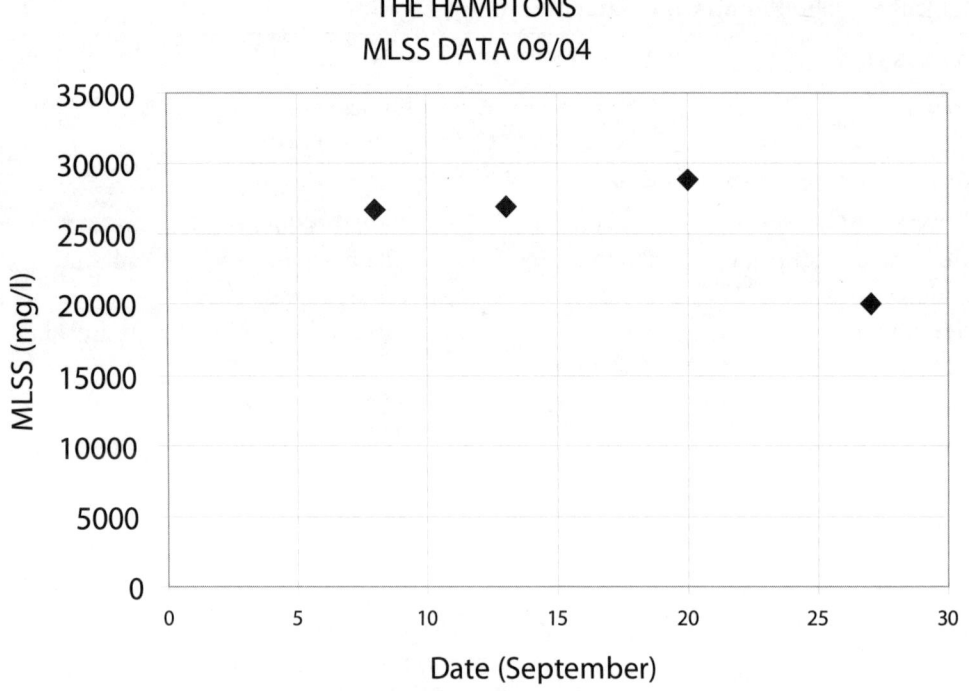

FIGURE 6.7—MIXED LIQUOR SUSPENDED SOLIDS DATA FROM THE HAMPTONS WRF (mg/L = ppm).

to sustain sufficient hydraulic performance. To date, the membranes have been cleaned in place using a dilute solution of bleach one time. The cleaning was part of a scheduled maintenance and was not required as a result of a TMP increase.

## PLANT DATA AND DESIGN SUMMARY

Table 6.6 shows a summary of current process operation parameters and design parameters for the membrane process. Current flows are considerably lower than one-quarter of the future design flow. Therefore, only one out of the four MBRs is in service at this time.

Table 6.7 summarizes treatment results and permit requirements for the Hamptons WRF. Data indicate that current and future permit limits are being met except for future limits for phosphate ($PO_4$-P).

Table 6.6 The Hamptons Membrane Filtration Process Design Parameters.

| PARAMETERS | UNITS | CURRENT DESIGN | FUTURE DESIGN |
|---|---|---|---|
| Annual average flow | mgd[a] | 0.225 | 0.90 |
| Maximum day average flow | mgd | 0.45 | 1.8 |
| Minimum day average process temperature | °F[b] | 60 | 60 |
| Maximum day average process temperature | °F | 90 | 90 |
| Annual average flux | gpd/sq ft[c] | 14.7[d] | 14.7 |
| Peak hour average flux | gpd/sq ft | 29.4[d] | 29.4 |
| Membrane cassette type | — | EK400 | EK400 |
| Number of cartridges per cassette | — | 200 | 200 |
| Surface area per cassette[e] | sq ft | 1291 | 1291 |
| Number of cassettes installed per train | — | 5 | 5 |
| Number of trains per MBR (membrane tank) | — | 2 | 2 |
| Number of MBRs (membrane tank) | — | | 4 |

[a] To convert from mgd to m³/d, multiply mgd by 3785.
[b] To convert from °F to °C, subtract 32 from °F and multiply by 0.556.
[c] To convert from gpd/sq ft to L/m²·h, multiply gpd/sq ft by 1.698.
[d] Hydraulic capacity at 15°C (59°F) listed. Process designed to treat 3407 m³/d (0.9 mgd) at a minimum temperature of 12°C (53.6°F).
[e] To convert from sq ft to m², multiply sq ft by 0.093.

## HIGHLIGHTS

We have learned the following from this case study:

1. Building in multiple process trains allows for easy expansion to meet future demands without building in membrane capacity now. At present, the plant is equipped to handle 852 m³/d (0.225 mgd) and sees flows up to 530 m³/d (0.14 mgd). Three other basins can be retrofitted to increase the plant capacity to 3407 m³/d (0.9 mgd).

2. Influent nitrogen can been reduced by approximately 93% from 56 mg/L (56 ppm) TKN to less than 4 mg/L (4 ppm) total nitrogen using a single recycle stream from the MBR to the AX basin.

Table 6.7 The Hamptons Treatment Results and Permit Requirements.

| PARAMETER[a] | AVERAGE INFLUENT | AVERAGE EFFLUENT[b] | CURRENT LIMITS[c] | FUTURE LIMITS[d] |
|---|---|---|---|---|
| $CBOD_5$ | 233 mg/L[e] | <2.0 mg/L[f] | Not permitted | <2 mg/L |
| TSS | 205 mg/L | <1.0 mg/L[f] | <5 mg/L | <1 mg/L |
| TKN | 56 mg/L | <1.0 mg/L[f] | Not permitted | Not permitted |
| Total nitrogen | Not available | <4.0 mg/L | Not permitted | <8 mg/L |
| TIN | Not available | Not available | Not permitted | Not permitted |
| $PO_4$-P | Not available | <5.0 mg/L | Not permitted | <0.13 mg/L |
| Turbidity | Not available | Not available | <3.0 NTU | <2.0 NTU |

[a] $CBOD_5$ = carbonaceous five-day biochemical oxygen demand; TSS = total suspended solids.
[b] Effluent values based on samples taken between November 2003 and August 2004 at a sample frequency of four per month.
[c] Plant currently operated under a land application systems (LAS) permit.
[d] Anticipated future limits.
[e] 1 mg/L = 1 ppm.
[f] Detection limits listed.

3. If necessary, the MBR MLSS concentration can be maintained at or above 25 000 mg/L (25 000 ppm) without significantly decreasing membrane performance.

## Case Study No. 3—Traverse City Regional Wastewater Treatment Plant

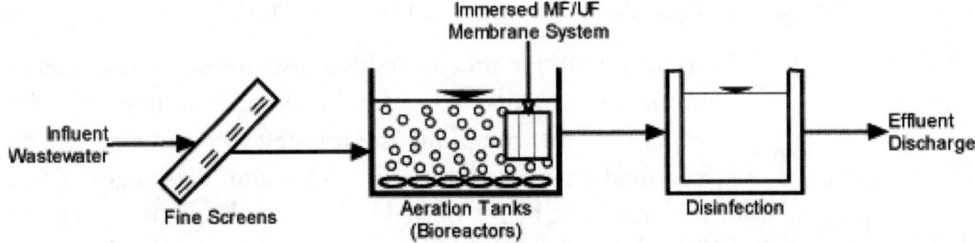

### BACKGROUND

Annual growth rates for Grand Traverse County, Michigan, will exert increasing demand on the area's wastewater treatment capacity. Before renovations, loadings to the Traverse City regional wastewater treatment plant (WTP) were projected to exceed the prior plant capacity by 2005. Capacity to treat a 40% increase in BOD load is expected to be required over the next 25 years.

Area residents expressed a desire that the city and townships make maximum use of the existing Traverse City WTP site (see Table 6.8) before building new collection systems or constructing a new treatment plant. Obstacles existed, however, to expansion of the existing plant by traditional methods, as the current site allows no opportunity to expand the plant footprint. The facility is bordered by railroad tracks to the north, Boardman Lake to the south, Boardman River to the west, and the recently developed Hull Park and regional library to the east. Additionally, the Michigan Department of Environmental Quality (DEQ) and local residents were decidedly uninterested in any increase in effluent loadings to surface waters—in fact, the community voluntarily decided to reduce effluent loadings to levels much lower than suggested by the DEQ. Consequently, any additional capacity for the existing plant would have to be achieved

Table 6.8  Traverse City Regional Wastewater Treatment Plant.

| POINTS OF INTEREST | DESCRIPTION | NOTES |
| --- | --- | --- |
| Location | Traverse City, Michigan | Facility is bordered on all sides, limiting the facility footprint |
| Year of system startup | 2004 | |
| Process configuration | MBR | |
| Membrane equipment configuration | ZeeWeed® 500c Vertical hollow fiber | |
| Permit type | Direct discharge | Issued by Michigan DEQ |
| Design flow/peak factor* | 10.5-mgd maximum monthly flow, 17-mgd peak hourly flow | |
| Unique features | As of 2004, the largest MBR plant in North America and the largest peak flow capacity in the world | The plant was designed as a full-flow membrane bioreactor plant with no upstream or in-tank equalization |

* 1 mgd = 3785 $m^3$/d.

by means of a space-efficient technology that increases plant capacity while producing a higher-quality treated effluent. Membrane bioreactor technology fit these requirements.

The completed facility is capable of treating maximum monthly loads of 9000 kg/d (20 000 lb/d) biochemical oxygen demand (BOD) at 32 000 m$^3$/d (8.5 mgd), with peak flows of 65 000 m$^3$/d (17.0 mgd). The facility design is in compliance with current Michigan DEQ requirements for wastewater treatment, including 25 mg/L (25 ppm) effluent BOD and 30 mg/L (30 ppm) effluent total suspended solids (TSS) on a monthly average basis. The current permit also establishes nutrient limits of 11 mg/L (11 ppm) for ammonia-nitrogen (NH$_3$-N) and an effluent phosphorus limit of 1 mg/L (1 ppm). To provide a desirable discharge to the natural environment, Traverse City has set voluntary target objectives of 4 mg/L (4 ppm) TSS, 4 mg/L (4 ppm) BOD, 1 mg/L (1 ppm) NH$_3$-N, and 0.5 mg/L (0.5 ppm) phosphorus in the effluent.

At the end of 2004, the Traverse City plant was the largest operating membrane bioreactor plant in North America based on design average flow and the largest in the world based on peak-flow capacity. The plant is also unique in that it was designed as a full-flow membrane bioreactor plant with no upstream or in-tank flow equalization.

## PROCESS OVERVIEW

Before the retrofit, primary treatment at the plant consisted of fine screens with 6-mm (0.24-in.) openings, followed by grit removal and primary clarification. Traveling band screens with 2-mm (0.08-in.) openings were added downstream of existing primary clarifiers to provide additional protection for the membranes.

The biological process design is based on full biological nutrient removal to remove required nitrogen and phosphorus constituents. The biological process is a variation of the University of Cape Town (UCT) process configuration, with adjustments made because of the highly oxygenated and nitrate-rich nature of the recirculating mixed liquor. As in the UCT process, the anoxic and anaerobic zones and recycles are configured to promote the formation of phosphorus-accumulating organisms and accomplish EBPR. existing anoxic and anaerobic zones, totaling 1949 m$^3$ (515 000 gal) and, therefore, 23% of the total bioreactor tank volume, were retained, as were the existing 4731 m$^3$ (1.25 mil. gal) of aerated zones. Eight membrane tanks were constructed to accommodate the membrane equipment, with a total additional volume of 1703 m$^3$ (450 000 gal). The biological process is provided in two parallel trains, with the capability to combine flows before the membrane separation stage or maintain separate trains throughout the membrane separation stage, as might be required for testing and optimization of the process.

The new membrane tank consists of eight equal compartments, each containing one-eighth of the membrane equipment (Figure 6.8). Channels and flow-splitting devices equally distribute mixed liquor to the eight compartments. The devices are arranged to allow any foam to migrate to the

FIGURE 6.8—THE TRAVERSE CITY MBR PLANT.

compartments rather than being trapped in channels, which may create a nuisance. A new building to house the pumping and equipment gallery is constructed integral to the membrane tanks. Two mixed liquor wasting pumps remove excess sludge solids on a regular basis for thickening and digestion.

Foam control is achieved by allowing it to travel through the eight compartments to the discharge-side mixed liquor channel. A foam pit is located at one end of the channel, with baffle separation to allow foam to overflow to the pit. The drain pump and the two wasting pumps have additional suction lines from the foam pit to allow the removal of foam either to the treatment process or to thickening and digestion.

## PLANT DATA AND DESIGN SUMMARY

Tables 6.9 and 6.10 summarize data from this case study.

## HIGHLIGHTS

We have learned the following from this case study:

1. Membranes treat the entire wastewater stream with no equalization.
2. In situ membrane cleaning can be provided without requiring removal of membranes.
3. The plant includes an innovative additional cleaning system, referred to as *empty-tank maintenance cleaning*.
4. Biological nutrient removal can be used in combination with membrane bioreactor technology.
5. Existing UV disinfection facilities can be modified to accommodate higher flows while recognizing the higher transmissivity of the effluent.

# Case Study No. 4—Key Colony Water Reuse Plant

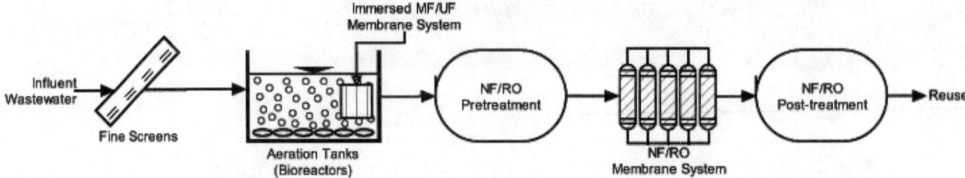

TABLE 6.9  Traverse City Membrane Filtration Process Design Parameters.

| Parameters | Units | Current Design |
|---|---|---|
| Annual average flow | mgd[a] | 8.5 |
| Peak hour flow | mgd | 17 |
| Maximum day average process temperature | °F[b] | 50 |
| Annual average flux | gpd/sq ft[c] | 9.7 |
| Peak hour flux | gpd/sq ft | 23.2 |
| Membrane cassette type | — | ZW-500c |
| Number of modules installed per cassette | — | 32 |
| Surface area per cassette[d] | ft² | 7040 |
| Number of cassettes installed per train | — | 13 |
| Number of membrane tanks (pumping trains) | — | 8 |

[a] To convert from mgd to m³/d multiply mgd by 3785.
[b] To convert from °F to °C; subtract 32 from °F and multiply by 0.556.
[c] To convert from gpd/sq ft to L/m²·h, multiply gpd/sq ft by 1.698.
[d] To convert from sq ft to m², multiply sq ft by 0.093.

Table 6.10  Traverse City Treatment Results and Permit Requirements.

| Parameter | Average Influent | Average Effluent | Current Limits | Voluntary Limits |
|---|---|---|---|---|
| $CBOD_5$[a] | 300 mg/L[b] | <4 mg/L | 25 mg/L | 4 mg/L |
| TSS | 517 mg/L | <4 mg/L | 30 mg/L | 4 mg/L |
| $NH_3$-N | 35 mg/L | <1 mg/L | 11 mg/L | 1 mg/L |
| Total phosphorus | 8 mg/L | <0.5 mg/L | 1 mg/L | 0.5 mg/L |
| Turbidity | >100 NTU | <0.5 NTU | — | — |

[a] $CBOD_5$ = carbonaceous five-day biochemical oxygen demand.
[b] 1 mg/L = 1 ppm.

# BACKGROUND

Key Colony, located in the Florida Keys, historically treated its municipal wastewater with a conventional activated sludge process (see Table 6.11).

Table 6.11 **Key Colony Water Reuse Plant (Courtesy of ZENON Environmental, Inc.).**

| POINTS OF INTEREST | DESCRIPTION | NOTES |
| --- | --- | --- |
| Location | City of Key Colony Beach, Florida | Located in the Florida Keys near fragile coral reefs |
| Year of system startup | 1999 | |
| Process configuration | MBR | |
| Membrane equipment configuration | ZeeWeed® 500a and ZeeWeed® 500c Vertical hollow fiber | ZENON |
| Permit type | MBR pretreatment for reverse osmosis | |
| Design flow/peak factor | 0.34/0.85 mgd* | |
| Unique features | The MBR is a retrofit of an existing steel package plant | The brackish MBR effluent is directly fed to a reverse-osmosis unit for water reuse |

* 1 mgd = 3785 m$^3$/d.

In 1998, the conventional system was nearing the end of its operating life and treatment capacity. In selecting a new treatment process, the city had several key factors to consider, including the protection of the local marine environment (coral reefs), the ability of the system to meet future discharge requirements, and the ease of increasing process capacity.

In 1999, the city awarded a contract to ZENON to install a ZeeWeed® MBR system. The ZeeWeed® MBR process demonstrated the ability to consistently achieve stringent discharge criteria for BOD, TSS, total nitrogen, and total phosphorus. This was very important, as the city was anticipating that these criteria would be added to the discharge permit in the near future.

In Key Colony, the wastewater is brackish because of high seawater infiltration and needs to be desalinated before the effluent can be used to irrigate a local golf course. A system was designed whereby 10% of the effluent of a ZeeWeed® MBR is further treated by RO.

This steel tank membrane bioreactor was built on the site of a former steel package plant, rehabilitating the former tanks and repairing/replacing the process equipment. The plant was designed to treat 1287 to 3217 m$^3$/d (0.34 to 0.85 mgd), depending on the amount of seasonal rainfall (i.e., peak rainfall during hurricane season), and has to meet Florida's 5-5-3-1 (milligrams per liter BOD-suspended solids-total nitrogen-total phosphorus) water-reuse criteria.

## PROCESS OVERVIEW

Figure 6.9 shows an overview of the process flow for the Key Colony water reuse plant (WRP).

The Key Colony WRP consists of two separate process trains, each with a dedicated set of ancillary equipment. The process trains are divided into three distinct zones—anoxic, aerated/anoxic, and aerobic—that are separated by concrete baffles.

Once the raw wastewater is prescreened, it is fed to the anoxic zone with an operating capacity of 208 m$^3$ (55 000 gal). Dissolved oxygen levels are maintained at less than 0.2 mg/L (0.23 ppm), and the majority of denitrification occurs in this zone. The mixed liquor then flows by gravity to zone two, where both nitrification and denitrification occur, minimizing the ammonia and nitrate concentrations in the aerobic zone. The ZeeWeed® immersed membranes are located in the aerobic zone (Figure 6.10). The membrane modules are connected to a permeate pump, which creates a low-pressure vacuum of –6.89 to –55.1 kPa (–1 to –8 psi). In an "outside-in" flow path, water is drawn from the mixed liquor through the surface of the membrane and into the fiber lumen. From there, the treated water is drawn out through the discharge system. Air is supplied to the membranes, where it emerges as a coarse bubble stream from the bottom of the membrane module. The air performs several key functions, including mixing of the biomass, membrane surface cleaning, and a portion of the process aeration. The immersed membranes can be operated effectively

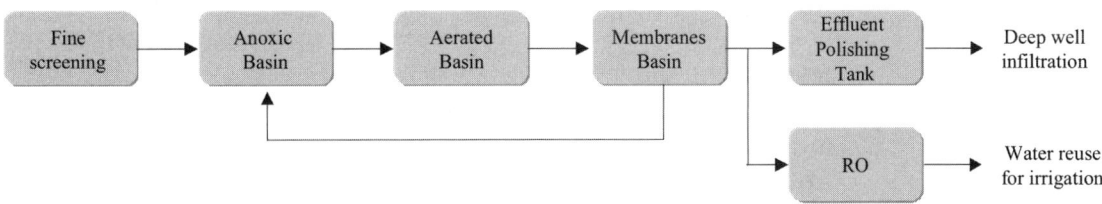

FIGURE 6.9—PROCESS FLOW DIAGRAM FOR THE KEY COLONY WATER REUSE PLANT (COURTESY OF ZENON ENVIRONMENTAL, INC.).

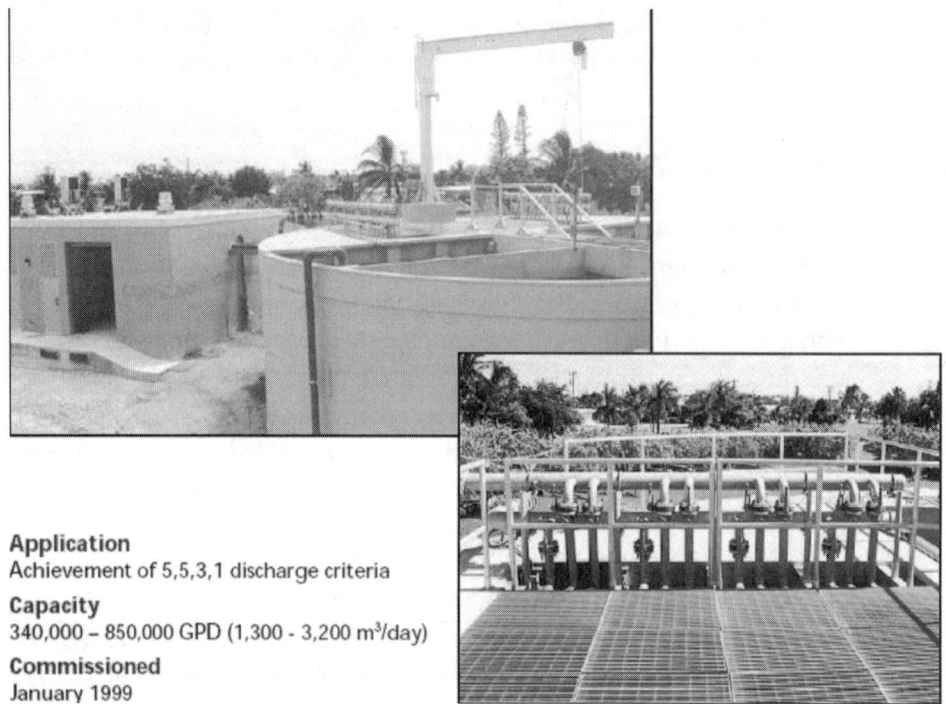

**Application**
Achievement of 5,5,3,1 discharge criteria
**Capacity**
340,000 – 850,000 GPD (1,300 - 3,200 m³/day)
**Commissioned**
January 1999

**FIGURE 6.10—KEY COLONY WATER REUSE PLANT (COURTESY OF ZENON ENVIRONMENTAL, INC.).**

at MLSS concentrations of up to 15 000 mg/L (15 000 ppm). The increased mass of biological solids per unit bioreactor volume ensures a larger population of nitrifiers, thus allows the ZeeWeed® MBR process to be operated at reduced organic loading rates and elevated SRTs (>15 days). Because the nitrifiers are trapped inside the bioreactor by the immersed membrane and are not washed away as in conventional plants, these bacteria show rapid metabolic recoveries in case of a toxic event.

A portion of the effluent from the ZeeWeed® MBR process is supplied to a local golf course for landscape irrigation, reducing the amount of fresh water used for this application. Before use at the golf course, the effluent is polished by a RO system. The effluent produced by the immersed membranes in the MBR is an ideal feed source for the RO membranes, as it is low in TSS and yields a silt density index (SDI) value <3.

Effluent that is not used for landscape irrigation is injected to deep wells and eventually flows back to the ocean.

## PLANT DATA AND DESIGN SUMMARY

Tables 6.12 and 6.13 show design information from this case study.

## HIGHLIGHTS

We can learn the following from this example:

1. The original externally fed drum screen was replaced with a 2-mm (0.08-in.) traveling band screen to provide more effective screening. The trash washer/compactor has worked very well.

Table 6.12  Key Colony Membrane Filtration Process Design Parameters (Courtesy of ZENON Environmental, Inc.).

| Parameters | Units | Current Design |
|---|---|---|
| Annual average flow | mgd[a] | 0.25 |
| Peak day average flow | mgd | 0.75 |
| Minimum day average process temperature | °F[b] | 77 |
| Maximum day average process temperature | °F | 86 |
| Annual average flux | gpd/sq ft[c] | 7.2 |
| Peak day average flux | gpd/sq ft | 21.7 |
| Membrane cassette type | — | ZW-500a/c |
| Number of modules installed per cassette | — | 500a: 8 <br> 500c: 22 |
| Surface area per cassette[d] | sq ft | 500a: 4000 <br> 500c: 4840 |
| Number of cassettes installed per train | — | 4 |
| Number of membrane tanks (pumping trains) | — | 2 |

[a] To convert from mgd to m$^3$/d multiply mgd by 3785.
[b] To convert from °F to °C, subtract 32 from °F and multiply by 0.556.
[c] To convert from gpd/sq ft to L/m$^2$·h, multiply gpd/sq ft by 1.698.
[d] To convert from sq ft to m$^2$, multiply sq ft by 0.093.

Table 6.13  Key Colony Treatment Results and Permit Requirements (Courtesy of ZE Environmental, Inc.).

| Parameter | Average Influent | Average Effluent | Current Permit Limits |
|---|---|---|---|
| CBOD$_5$[a] | 220 mg/L[b] | <2 mg/L | 5 mg/L |
| TSS | 150 mg/L | <5 mg/L | 5 mg/L |
| TKN | 40 mg/L | — | — |
| NH$_3$-N | 31 mg/L | <1 mg/L | — |
| Total phosphorus | 5.2 mg/L | <1 (demonstrated—not currently dosing coagulant) | 1 mg/L |
| Total nitrogen | — | <3 (demonstrated—not currently required) | 3 mg/L |

[a] CBOD$_5$ = carbonaceous five-day biochemical oxygen demand.
[b] 1 mg/L = 1 ppm.

2. Cyclic aeration was added after one year of operation and resulted in 50% reduction of energy consumption.
3. The MBR provides a high-quality effluent for RO membranes and eliminates the fouling caused by suspended solids that is associated with conventional pretreatment.

## Case Study No. 5—West Basin Water Recycling Plant

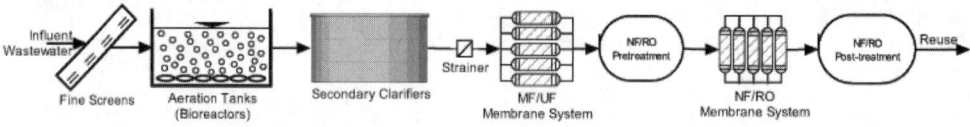

### BACKGROUND

Severe droughts between the late 1980s and early 1990s forced the West Basin Municipal Water District in Los Angeles, California (see Table 6.14), to examine and improve its management of resources and recycling/conservation efforts to ensure a reliable supply of water. State and federal

Table 6.14 West Basin Water Recycling Plant (WBWRP).

| POINTS OF INTEREST | DESCRIPTION | NOTES |
|---|---|---|
| Location | Los Angeles, California | Influent for WBWRP is the effluent from the nearby Hyperion Treatment Plant |
| Year of system startup | 1995 | |
| Process configuration | MF pretreatment: Reverse osmosis | |
| Membrane equipment configuration | USFilter/Memcor | |
| Permit type | | |
| Design flow/peak factor[a] | 12 mgd* | Build-out capacity of 100 mgd |
| Unique features | Highly flexible process | Allows for meeting of a variety of water quality/reuse requirements |

* 1 mgd = 3785 m$^3$/d.

funding was granted to the District in 1992 to pursue its water-recycling program. In three years, the district completed the construction of the West Basin water recycling plant (WBWRP), which was placed online in February 1995. The WBWRP uses two distinct treatment processes, a conventional direct filtration system and a membrane treatment system, to produce nonpotable recycled water for a wide range of beneficial uses. Currently, the operating capacity of the WBWRP is $1.61 \times 10^5$ m$^3$/d (42.5 mgd), with an ultimate build-out capacity of $3.79 \times 10^5$ m$^3$/d (100 mgd). The current installed membrane capacity at the WBWRP is approximately $4.54 \times 10^4$ m$^3$/d (12 mgd).

The source water for the WBWRP is secondary effluent from the City of Los Angeles Hyperion Treatment Plant (HTP). The HTP is a $1.7 \times 10^6$-m$^3$/d (450-mgd) wastewater treatment facility consisting of screening facilities and grit removal, primary settling tanks, high-purity oxygen reactors, and final clarification. A $2.16 \times 10^5$-m$^3$/d (57-mgd) secondary effluent pumping station transfers effluent from the HTP to the WBWRP.

## PROCESS OVERVIEW

As previously stated, the WBWRP uses conventional direct filtration and membrane treatment systems to produce nonpotable recycled water for a variety of uses, including irrigation, industrial cooling and process water, and seawater intrusion barriers. Each use demands a unique finished water quality, which requires a unique process train. Conventional direct filtration and membrane treatment processes provide the foundation for the WBWRP's four distinct process trains. The first process train consists of flocculation, direct filtration, and disinfection, and the effluent is used to produce water suitable for irrigation demands. Additional offsite treatment, including nitrification and denitrification, is used to further process a portion of this water for use in cooling towers. The total capacity of the conventional direct filtration train is $1.14 \times 10^5$ m³/d (30 mgd); however, current upgrades will increase the treatment capacity of this train to $1.51 \times 10^5$ m³/d (40 mgd). The second process train is composed of decarbonation, lime clarification, filtration, and single-pass RO treatment to produce $1.9 \times 10^4$ m³/d (5 mgd) of water suitable for use as a seawater intrusion barrier. The third train produces an additional 9463 m³/d (2.5 mgd) of barrier water through the use of automatic strainers, MF, cartridge filtration, chemical pretreatment, and single-pass RO. The fourth process train uses similar technology to the third train and consists of automatic strainers, MF, cartridge filtration, chemical pretreatment, and double-pass RO treatment to produce $1.63 \times 10^4$ m³/d (4.3 mgd) of water suitable for boiler feed systems. The use of a double-pass RO system produces extremely high-quality permeate. Water produced from the first pass contains less than 60 mg/L (60 ppm) of total dissolved solids and is suitable for use in low-pressure boilers. Product water from the second pass contains less than 5 mg/L (5 ppm) of total dissolved solids and is suitable for use in high-pressure boilers. The total capacity of the membrane treatment systems is $4.73 \times 10^4$ m³/d (12.5 mgd); however, current planning will increase the treatment capacity of these trains to $8.52 \times 10^4$ m³/d (22.5 mgd).

Prolonged side-by-side operation of the second and third treatment trains has revealed the following:

- Microfiltration produces a higher-quality (in terms of turbidity and SDI) and more consistent water than lime clarification.
- Maintenance requirements are significantly less for the MF/RO train than for the lime clarification/RO train.

- Chemical cleaning frequency for the RO membranes fed by microfiltered product water is approximately every 6 to 12 months.
- Chemical cleaning frequency for the RO membranes fed by lime-clarified product water is as frequent as once every month.

Microfiltration/Double-Pass Reverse-Osmosis Process Description

The first step in MF/double-pass RO is to strain the secondary effluent feed stream using 500-$\mu$m (0.02-in.) automatic strainers. These strainers use a stationary basket and a rotating arm to filter the secondary effluent feed stream. Automatic cleanings are initiated either by a predetermined time interval or by a predetermined pressure drop.

Microfiltration is accomplished by ten banks of outside-in, 0.2-$\mu$m (8 × $10^{-6}$-in.) polypropylene hollow-fiber membranes manufactured by USFilter/Memcor. Each bank contains 90 MF modules, each containing 34.2 $m^2$ (368 sq ft) of surface area. Operating conditions of the MF system are a flux of approximately 34 L/$m^2 \cdot$h (20 gpd/sq ft) and a pressure drop of approximately 137.9 to 172.4 kPa (20 to 25 psi). Productivity of the MF system is maintained through periodic chemical cleanings and maintaining a chloramine residual through the system. Chemical cleanings, using citric acid, sodium hydroxide, and surfactants, are required approximately once every three weeks. Hypochlorite is also added to the feed stream at a concentration of approximately 3 to 5 mg/L (3 to 5 ppm) to promote the formation of a chloramine residual. The chloramine residual provides protection from biofouling for both the MF system and the RO system downstream.

Operation of a RO system at elevated recoveries often results in the supersaturation of sparingly soluble inorganic salts. The addition of a threshold inhibitor minimizes the potential for these salts to precipitate out of solution and foul the membranes. Sulfuric acid is used to maintain the desired equilibrium between carbonate and bicarbonate species in RO feed and concentrate streams. The pretreatment chemicals are added upstream of cartridge filters to remove any impurities before the RO units.

Reverse-osmosis pretreatment is accomplished via cartridge filters with a 20-$\mu$m (8 × $10^{-4}$-in.) rating. While the pore size of these cartridge filters is one order of magnitude larger than that of the upstream microfilters, cartridge filtration is still used to protect the RO membranes should the integrity of the MF system be compromised.

Final treatment is accomplished via a two-pass RO membrane system. Both the first- and second-pass RO units use the ESPA2 spiral-wound membrane elements manufactured by Nitto Denko/Hydranautics. Each ESPA2 element contains 37.2 m$^2$ (400 sq ft) of available surface area and is composed of a polyamide membrane barrier supported by a polysulfone structure, which provides superior rejection, fouling resistance, and chemical tolerance than cellulose acetate membranes. The first-pass RO system contains two units (one online and one standby), each producing a maximum of 8706 m$^3$/d (2.3 mgd). Each unit contains 72 pressure vessels with seven elements in each vessel. The first-pass RO system operates at a flux of approximately 19.4 L/m$^2$·h (11.4 gpd/sq ft) and requires a chemical cleaning approximately once or twice each year. The second-pass RO system contains three units (two online and one standby), each producing a maximum of 4921 m$^3$/d (1.3 mgd). Each unit contains 24 pressure vessels with seven elements in each pressure vessel. The second-pass system operates at a flux of approximately 32.8 L/m$^2$·h (19.3 gpd/sq ft) and requires a chemical cleaning approximately once or twice each year.

Product Water Quality

The MF/double-pass RO system produces $1.89 \times 10^4$ m$^3$/d (5 mgd) of water for both low- and high-pressure boilers, each with unique feed water quality requirements.

## PLANT DATA AND DESIGN SUMMARY

Tables 6.15 and 6.16 provide information from this case study.

## HIGHLIGHTS

The following summarizes what we have learned from this example:

1. Membrane technology is capable of treating wastewater effluent to produce product water of varying qualities that is suitable for a wide range of applications, including irrigation, industrial cooling and process water, low- and high-pressure boiler feed, and groundwater barriers to seawater intrusion.

Table 6.15  Effluent Water Quality from the Hyperion Treatment Plant (HTP).

| Parameter | Units | HTP Effluent |
|---|---|---|
| Total dissolved solids | mg/L* | 773 |
| Bicarbonate | mg/L as $CaCO_3$ | 293 |
| Calcium | mg/L | 48 |
| Silica | mg/L | 26.5 |
| Total organic carbon | mg/L | 11 |
| BOD | mg/L | 28 |
| Suspended solids | mg/L | 11 |
| Ammonia | mg/L as N | 24.4 |
| Turbidity | NTU | 5.6 |

* 1 mg/L = 1 ppm.

Table 6.16  First- and Second-Pass RO Product Water Quality.

| Parameter | Units | First Pass | Second Pass |
|---|---|---|---|
| Sulfate | mg/L* | ND | ND |
| Chloride | mg/L | 2.3 | ND |
| Nitrate (as N) | mg/L | ND | ND |
| Nitrite (as N) | mg/L | ND | ND |
| Calcium | mg/L | 0.06 | ND |
| Magnesium | mg/L | 0.02 | ND |
| Hardness (as $CaCO_3$) | mg/L | 0.23 | ND |
| Sodium | mg/L | 4.25 | 0.25 |
| Potassium | mg/L | 0.31 | ND |
| Silica | mg/L | 0.19 | ND |
| Total organic carbon | mg/L | 0.5 | 0.2 |
| Total dissolved solids | mg/L | 14 | 2 |

*1 mg/L = 1 ppm.

2. Microfiltration technology provided superior pretreatment for subsequent RO processes compared with conventional lime clarification. Pretreatment evaluations were based on both maintenance and chemical cleaning frequency of the RO membranes.

3. Biofouling of both the MF and RO systems was controlled through the use of a chloramine residual. Chloramine formation was achieved through the addition of hypochlorite.
4. A MF/double-pass RO treatment train is capable of producing very high-purity water suitable for various industrial applications.

## Case Study No. 6—Chandler, Arizona, Reverse-Osmosis Wastewater Reclamation Plant

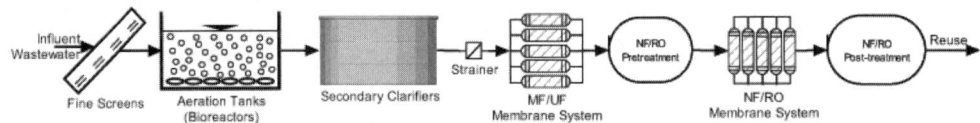

### BACKGROUND

In 1996 Chandler, Arizona, commenced operation of a membrane-based facility (see Table 6.17) that treats wastewater from a semiconductor manufacturing plant to recharge an aquifer with high-quality reclaimed water to improve the area's overall water balance and, at the same time, reduce the volume of wastewater treated at the city's conventional wastewater treatment plant. The city of Chandler worked in partnership with Intel Corporation to develop this project to attract one of the world's largest semiconductor plants and the associated thousands of technology jobs to the desert community on the south side of the Phoenix metropolitan area. According to then Chandler Director of Public Works, George Selvia, this project is a "positive win-win situation for provision of water in our desert climate!" An overview of the wastewater treatment plant is shown in Figure 6.11.

As shown on the overview figure, the semiconductor plant is supplied with water from the city's distribution system. Depending on the time of year, the water originates as either surface or groundwater. A variety of water-conservation methods are used at the semiconductor "Fab" facility, including supply to cooling towers and landscape irrigation from conventionally reclaimed municipal effluent. Wastewater generated at the

Table 6.17 Chandler Industrial-Municipal Water Reclamation Plant.

| Points of Interest | Description | Notes |
|---|---|---|
| Location | Chandler, Arizona | Phoenix metropolitan area |
| Year of system startup | 1996 | MF added in 1999 |
| Process configuration | MF pretreatment for RO and NF on a sidestream | MF—Pall/Asahi PVDF<br>RO—Dow BW30 polyamide<br>NF—Dow NF70 polyamide |
| Membrane equipment configuration | Hollow-fiber MF and spiral-wound RO and NF | RO—5 trains (25% duty each)<br>NF—3 trains (50% duty each) |
| Permit type | Aquifer protection | |
| Design flow/peak factor[a] | 2 mgd* | Reclaims approximately 80% of the source water |
| Unique features | Industrial–municipal partnership | The City of Chandler and the Intel Corporation partnered on this original concept |

* 1 mgd = 3785 m³/d.

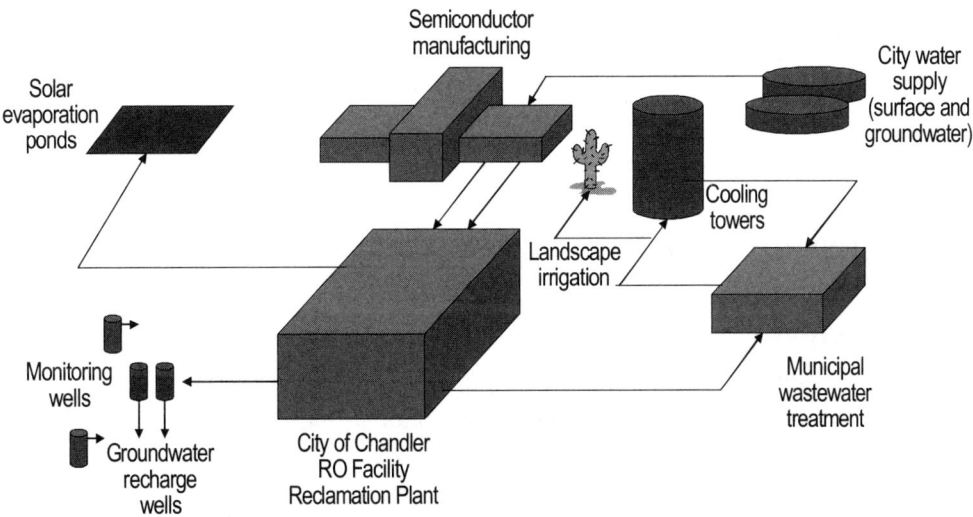

FIGURE 6.11—OVERVIEW OF CHANDLER WASTEWATER RECLAMATION PLANT (FREEMAN ET AL., 2005) (COURTESY OF AMERICAN WATER WORKS ASSOCIATION).

industrial plant is segregated into three streams: standard municipal effluent and two separate process streams that flow to the Chandler RO reclamation plant (RORP). Approximately 80% of the RORP influent is reclaimed for aquifer recharge. The remaining 20% is discharged to solar evaporation ponds or other municipal wastewater treatment plants, depending on seasonal needs. Overall, the Fab campus was designed for a net water usage of half the daily volume, 7570 m$^3$/d (2 mgd), the amount that a residential subdivision occupying the same land area would consume. The Chandler RORP has recharged the aquifer with 9.46 mil. m$^3$ (2.5 bil. gal) between startup in mid-1996 and the end of 2004.

Two separate sets of processes provide source water to the Chandler RORP. The quantity and flowrate to the RO100 part of the RORP changes frequently, even varying multiple times a day. The total dissolved solids concentration ranges from approximately 280 to 1200 mg/L (280 to 1200 ppm); the pH is typically in the 6 to 10 unit range; and the temperature varies seasonally in the 22° to 30° C (71.6° to 86° F) range. The influent to RO200 is more stable. There is some seasonal variation to the RO200 influent, but the daily range is generally less pronounced. The RO200 influent exhibits a total dissolved solids range of 1500 to 2500 mg/L (1500 to 2500 ppm), a pH range of 7 to 9, and a temperature range of 22° to 30° C (71.6° to 86° F). Initial feasibility studies for the project called for segregation of RO100 and RO200 influent to avoid calcium fluoride scaling in the RO equipment; however, actual water quality subsequently indicated that the segregation was not necessary.

## PROCESS OVERVIEW

The Chandler RORP applies a triple-membrane process, which includes MF, RO, and NF to produce drinking water quality reclaimed water for groundwater recharge (see Figure 6.12). The RO100 process line consists of a disinfectant addition to maintain a chloramine residual, two-stage pH-adjustment in which either caustic or sulfuric acid is automatically added depending on the influent pH, antiscalant addition to protect the RO from inorganic scale caused by concentration of sparingly soluble salts, MF to protect the RO from particles and lower the SDI below the desired 3 to 5 units, an additional dose of hypochlorite to maintain a chloramine

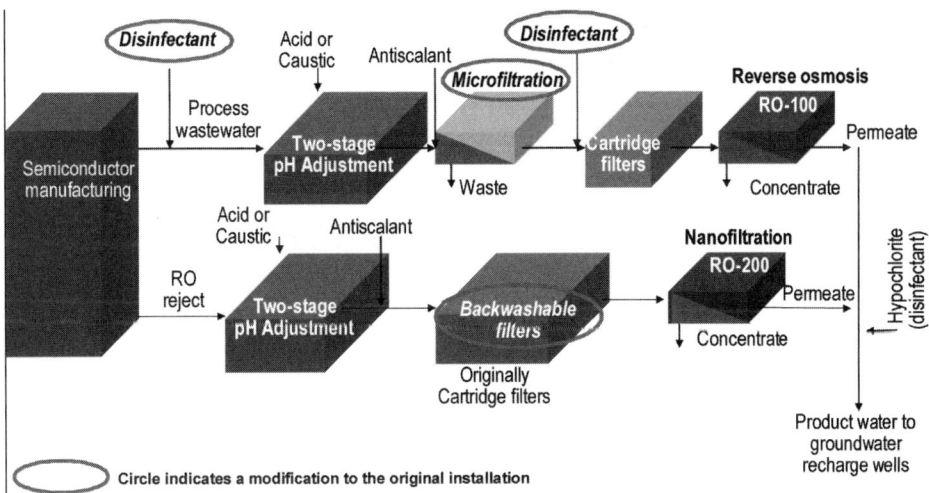

FIGURE 6.12—PROCESS FLOW DIAGRAM FOR THE CHANDLER WASTEWATER RECLAMATION PLANT (FREEMAN ET AL., 2005) (COURTESY OF AMERICAN WATER WORKS ASSOCIATION).

residual that helps control biofouling in the RO, 5-μm cartridge filtration, and the RO step. The RO200 process line consists of two-stage pH adjustment in which either caustic or sulfuric acid can be added depending on the influent pH, an antiscalant addition to protect the NF from inorganic scale caused by concentration of sparingly soluble salts, backwashable screen filters (rated at 10 to 12 μm) to protect the NF from particles, and the NF step. The following were added to the original 1996 facility: MF in 1999, disinfection in 2001, a pre-NF backwashable screen in 2003, and a single-element cleaning stand in 2003.

The reclamation plant was designed before the Fab facility was in operation, so site-specific, water-quality data were not available. At startup, it was found that additional pretreatment would be needed for the RO100 process line to protect the RO from high loadings of particles as indicated by high levels of SDI and turbidity. For RO, values of 3 to 5 SDI units and turbidity less than 1 NTU are preferred; at Chandler the SDI was too high to measure and the turbidity exceeded 60 NTU. As a short-term solution, field modifications were implemented to allow operation of the plant.

These modifications included lower flux, lower recovery, increased velocity across the membrane surface, lower feed pH to address calcium phosphate scaling, frequent cleanings, and experimental cleanings. Subsequently, UF prefiltration was tested, which was later converted to MF.

## PLANT DATA AND DESIGN SUMMARY

Tables 6.18, 6.19, and 6.20 show data derived from this case study.

Table 6.18 Chandler RO Facility Effluent Water Quality (Freeman et al., 2005) (Courtesy of American Water Works Association).

| Parameter[a] | Average of Injected Water (Aug 1996– Oct 2004) | 90th-Percentile of Injected Water (Aug 1996– Oct 2004) | U.S. Environmental Protection Agency Drinking Water Standards[b] | Is Injected Water Quality Less Than or Equal to Drinking Water Standards? |
|---|---|---|---|---|
| *Physical* | | | | |
| Conductance (field), μS/cm | 704 | 840 | NA | NA |
| Total dissolved solids | 292 | 424 | 500* | Yes |
| *Inorganic* | | | | |
| Alkalinity, total | 101 | 157 | NA | NA |
| Aluminum | <0.1 | <0.1 | 0.05 to 0.2 | Yes |
| Antimony | <0.003 | <0.003 | 0.006 | Yes |
| Arsenic | <0.005 | <0.005 | 0.010 (in 2006) | Yes |
| Barium | <0.01 | 0.08 | 2 | Yes |
| Beryllium | <0.001 | <0.001 | 0.004 | Yes |
| Boron | 0.28 | 0.40 | 1[b] | Yes |
| Cadmium | <0.001 | <0.001 | 0.005 | Yes |
| Calcium | 7.2 | 15 | NA | NA |

Table 6.18 Chandler RO Facility Effluent Water Quality (Freeman et al., 2005) (Courtesy of American Water Works Association) *(continued)*.

| PARAMETER[a] | AVERAGE OF INJECTED WATER (AUG 1996– OCT 2004) | 90TH-PERCENTILE OF INJECTED WATER (AUG 1996– OCT 2004) | U.S. ENVIRONMENTAL PROTECTION AGENCY DRINKING WATER STANDARDS[b] | IS INJECTED WATER QUALITY LESS THAN OR EQUAL TO DRINKING WATER STANDARDS? |
|---|---|---|---|---|
| Chloride | 100 | 177 | 250[b] | Yes |
| Chromium (total) | <0.01 | <0.01 | 0.1 | Yes |
| Magnesium | 2.7 | 6.2 | NA | NA |
| Mercury | <0.0002 | <0.0002 | 0.002 | Yes |
| Potassium | 2.3 | 3.7 | NA | NA |
| Selenium | <0.005 | <0.005 | 0.05 | Yes |
| Silica | 2.4 | 3.3 | NA | NA |
| Silver | <0.04 | <0.04 | 0.1 | Yes |
| Sodium | 107 | 146 | NA | NA |
| Strontium | <0.01 | <0.01 | NA | NA |
| Sulfate | 32 | 60 | 250 | Yes |
| *Nitrogen* | | | | |
| Nitrate (as N) | <2 | 3.3 | 10 | Yes |
| Nitrite (as N) | 0.23 | 0.29 | 1 | Yes |
| TKN (as N) | 0.83 | 1.8 | NA | NA |
| *Organic* | | | | |
| Total organic carbon | 1.1 | 2.4 | NA | NA |
| Benzene | <0.002 | <0.002 | 0.005 | Yes |
| Ethyl benzene | <0.002 | <0.002 | 0.7 | Yes |
| Toluene | <0.002 | <0.002 | 1 | Yes |
| Xylene, total | <0.004 | <0.004 | 10 | Yes |

[a] Concentrations in milligrams per liter (parts per million), unless otherwise noted.

[b] Indicates secondary standard or guideline.

Table 6.19 Chandler Field Modifications (Freeman et al., 2005) (Courtesy of American Water Works Association).

| Parameter | Target | Discussion |
| --- | --- | --- |
| Low flux | 8.5 to 9 gpd/sq ft[a] | Low flux selected to retard fouling by particles and colloids indicated by high SDI. |
| Low recovery | 80% (system recovery) | Initial design was based on very low total dissolved solids water with low risk of inorganic scaling. Actual analyses and calculations indicated 80% recovery achievable. |
| High velocity | >61 L/s (>16 gpm) per 0.2-m-diam (8-in.-diam) vessel | The installed equipment allowed recirculation of a portion of the concentrate, thereby allowing control of the concentrate flowrate exiting the vessels. Higher velocity helps to retard particulate fouling. |
| Adjusted pH | 5 pH units | Calcium and phosphate concentrations were higher than anticipated. On-site jar testing indicated scaling not a problem if feed adjusted to pH 5. |
| Cleaning | Frequency and type | Chemical cleaning every 1 to 4 weeks with limited effectiveness, eventually with the high pH–hypochlorite method described below. |
| Pretreatment | Options considered | Discussed in the next section. |

[a] To convert from gpd/sq ft to L/m²·h, multiply gpd/sq ft by 1.698.

# HIGHLIGHTS

The following describes some lessons learned from this example:

1. The RO- and NF-based, water-reclamation facility has consistently provided excellent finished water quality, even before plant modifications that better equipped the plant to hydraulically treat the challenging raw water.

2. From July 1996 to December 2004, more than 9.46 mil. m³ (2.5 bil. gal) of water that meets drinking water standards has been injected.

3. Field modifications enabled operations during the 18 months of operations while treating water outside the typical feed water requirements for RO of less than 3 SDI units and less than 1 NTU.

Table 6.20  Typical Chandler Operating Conditions in 2004 (Freeman et al, 2005) (Courtesy of American Water Works Association).

| DESCRIPTION | MF | RO (RO100) | NF (RO200) |
|---|---|---|---|
| Membrane type | PVDF, encased hollow fiber | Polyamide, spiral wound | Polyamide, spiral wound |
| Flux, gpd/sq ft[a] | 12 | 9.5 | 10 |
| Other conditions | 30-second filtered water backwash every 30 minutes and a 25-second air scrub every 4 hours | 80 to 88% recovery | 75% recovery |
| Transmembrane pressure, psi[b] | 4 to 10 | 90 to 150 | 125 to 150 |
| Cleaning frequency, months | 12[c] | 12 | 2 |
| Anticipated membrane life, years | 10 | 2.5 | 7 |

[a] To convert from gpd/sq ft to L/m²·h, multiply gpd/sq ft by 1.698.

[b] To convert from psi to kPa, multiply psi by 6.89.

[c] MF cleaning approximately every 12 months is not a result of fouling, but when that operation fits into the plant's maintenance cycle.

The main field modifications were low flux, low recovery, high velocity, low feed pH, and an experimental cleaning at high pH in the presence of hypochlorite.

4. The experimental cleaning method with pH of 11.8 to 12.0 and 50 to 80 mg/L (50 to 80 ppm) free chlorine was relatively successful at cleaning with inexpensive chemicals. At elevated pH, the hypochlorite is less aggressive as a disinfectant but also less aggressive as an oxidant of the chlorine-sensitive polyamide RO membrane. The operators had to carefully conduct these cleanings to prevent membrane damage.

5. Significant long-term modifications to the facility included adding pretreatment steps of MF and chloramination, as well as a lead element test stand.

# Case Study No. 7—Scottsdale Water Campus

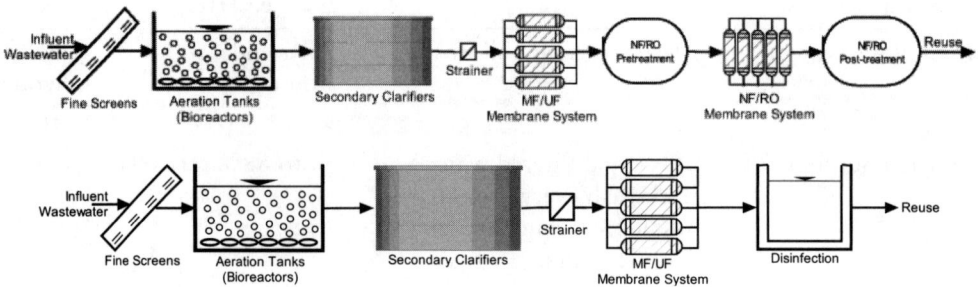

## BACKGROUND

The 1980 Arizona Groundwater Management Act was legislated to eliminate groundwater mining in active management areas (AMAs). In response, many communities have turned to artificial groundwater recharge as a way to beneficially reuse reclaimed water. Groundwater can continue to be withdrawn in AMAs in quantities equal to the artificial recharge water "bank account". Scottsdale is located within the Phoenix AMA and has limited access to surface water supplies. In 1999, the city implemented a large-scale, $4.54 \times 10^4$-m$^3$/d (12-mgd) water-reclamation facility (Water Campus; see Table 6.21). In the summer, reclaimed water is used to irrigate local golf courses, directly offsetting potable water demand. In the winter when demand is low, reclaimed water is treated with MF and RO before groundwater recharge. Figure 6.13 shows the process flow for the Scottsdale Water Campus.

The source water for the Water Campus is wastewater generated in Scottsdale. Source water for the membrane advanced water treatment facility (AWTF) is filtered secondary effluent. Influent water quality to the AWTF is presented in Table 6.22.

## PROCESS OVERVIEW

The Scottsdale Water Campus AWTF uses MF followed by RO to produce drinking water quality reclaimed water for groundwater recharge. Cartridge filters are included after MF as an added protection for RO membranes.

Table 6.21 Scottsdale Water Campus.

| Points of Interest | Description | Notes |
|---|---|---|
| Location | Scottsdale, Arizona | Phoenix metropolitan area |
| Year of system startup | 1999 | Expanded in 2002 |
| Process configuration | Microfiltration or MF pretreatment for RO | MF–USFilter-Memcor RO–Koch/Fluid Systems |
| Membrane equipment configuration | Memcor 90 M10C Koch Model 8832HR | 24 MF units and 14 RO trains |
| Permit type | Aquifer protection | |
| Design flow/peak factor | 12 mgd* | Expanded to 15 mgd in 2002 |
| Unique features | Part of an AMA strategy | First full-scale application of MF and polyamide thin film RO Overall facility treats wastewater to reclaimed water |

* 1 mgd = 3785 m³/d.

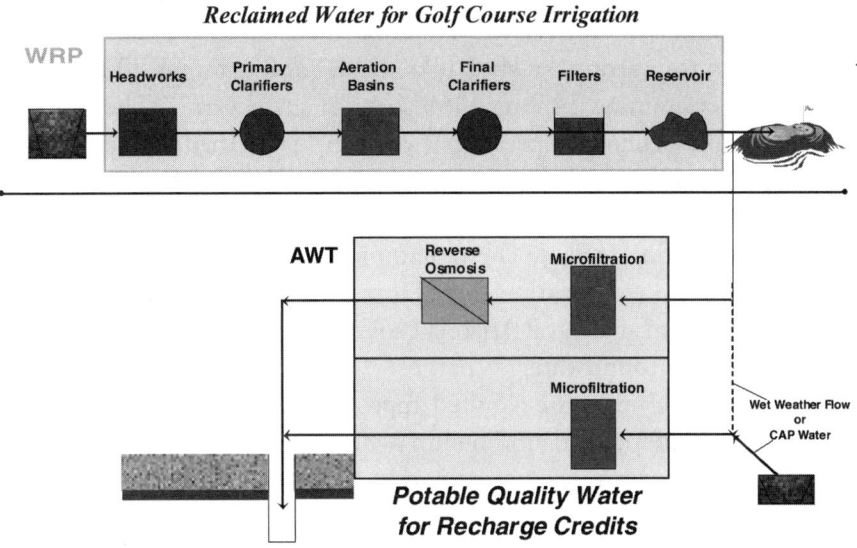

FIGURE 6.13—PROCESS FLOW DIAGRAM FOR THE SCOTTSDALE WATER CAMPUS (COURTESY OF BLACK AND VEATCH).

Table 6.22  Advanced Water Treatment Facility Influent Water Quality (Courtesy of Black and Veatch Corporation).

| Parameter | Units | Values |
|---|---|---|
| $BOD_5$ | mg/L[a] | < 10 |
| Turbidity | NTU | <2 |
| Total nitrogen | mg/L as N | < 10 |
| pH | | 6–9 |
| Fecal coliform | CFU/100 mL | Nondetectable |
| Total dissolved solids[b] | mg/L | 800–1200 |

[a] mg/L = ppm.

[b] Total dissolved solids are not controlled in the secondary effluent. Typical range of values is listed here.

Microfiltration

Water reclamation facility tertiary effluent is passed through 500-$\mu$m automatic backwash strainers before MF. Microfiltration is provided with 24 outside-in, 0.2-$\mu$m (8 × 10$^{-6}$-in.), polypropylene hollow-fiber membrane units manufactured by USFilter–Memcor 90 M10c. The units are arranged in four groups of six banks. Each bank contains 90 MF modules, each containing 34.2 m$^2$ (368 sq ft) of surface area. The installed capacity with all units in operation is 7570 m$^3$/d (2.0 mgd) operating at an average flux of 84.9 L/m$^2$·h (50 gpd/sq ft). One-half of the units' capacity, 3.78 × 10$^4$ m$^3$/d (10 mgd), treats excess dry-weather flow from the water reclamation facility. The remaining units comprise a parallel system for treating wet-weather flow. During dry weather, these units can also be used to treat Central Arizona Project water (Colorado River water) for additional groundwater recharge.

Chemical cleanings are required approximately once every two weeks. Sodium hypochlorite and ammonium hydroxide are also added to the MF feed stream to provide a chloramine residual. The chloramine residual provides protection from biofouling for both the MF and downstream RO.

Reverse Osmosis

Microfiltration effluent is passed through 5-$\mu$m cartridge filters before RO as an additional barrier of protection for the RO membranes. Although the cartridge filters have larger openings than the MF membranes, they are provided to protect the RO membranes if the conveyance system between MF and RO is compromised.

The RO membranes are configured in three-stage units. Ten trains were installed in 1999 and four in 2002. The 14 trains provide a feed water treatment capacity of $5.3 \times 10^4$ m$^3$/d, or 3785 m$^3$/d each(14 mgd, or 1 mgd each). Each of the 1999 trains is arranged in a 24:10:5 array, and the 2002 trains apply 20:10:5 arrays. Pretreatment includes addition of sulfuric acid to lower the pH to reduce calcium carbonate scaling potential and a threshold inhibitor, Hyposperse, to inhibit the precipitation of sparingly soluble salts. The RO membrane elements are thin-film composite polyamide manufactured by Koch/Fluid Systems, Model 8832HR.

The facility has an ongoing research program. For one study, a train is being operated at a flux of 21.2 L/m$^2$·h (12.5 gpd/sq ft) rather than the typical 18.2 L/m$^2$·h (10.7 gpd/sq ft). Large, 0.41-m-diam (16-in.-diam) elements are being evaluated in another train.

A permanently piped RO membrane cleaning system is provided for chemical cleaning of the RO membrane in situ. Chemical cleaning is accomplished by recirculating a chemical solution through each array of an offline train. The RO membranes have required cleaning less than once per year over the first 6 years of operations. The plant staff project element life at 10 to 14 years. This is a longer life than observed at many RO facilities and is probably a result of the high-quality prefiltration provided by the MF and maintenance of a chloramine residual that discourages biofouling.

Posttreatment Facilities

The posttreatment system consists of decarbonation and lime addition. The system is required to provide a stable water quality that will not be corrosive to the distribution system and will not leach the soil to which it is injected. The decarbonator will reduce the carbon dioxide ($CO_2$) concentration to approximately 8 mg/L (8 ppm) in the product. This $CO_2$ reduction will reduce the lime requirements.

Lime is added to stabilize the AWTF product water. Hydrated lime is received in bulk shipment as a dry powder or crystals and stored in two storage silos. Volumetric feeders feed a lime solution to the product water stream in a flow channel below the lime silos.

## PLANT DATA

Product water quality produced by the MF and RO systems is presented in Table 6.23.

Table 6.23  Advanced Water Treatment Facility Effluent Water Quality (Courtesy of Black and Veatch Corporation).

| Parameter | Units | Value |
|---|---|---|
| pH |  | 6.5–8.5 |
| Turbidity | NTU | 0.1 |
| Total dissolved solids | mg/L* | <280 |
| Chloride | mg/L | <250 |
| Fluoride | mg/L | <2 |
| Total nitrogen | mg/L | <5 |
| Sodium | mg/L | <100 |
| Sulfate | mg/L | <100 |
| Arsenic | mg/L | <0.002 |
| Barium | mg/L | <1 |
| Cadmium | mg/L | <0.01 |
| Chromium | mg/L | <0.05 |
| Copper | mg/L | <1 |
| Iron | mg/L | <0.3 |
| Lead | mg/L | <0.015 |
| Mercury | mg/L | <0.002 |
| Selenium | mg/L | <0.01 |
| Silver | mg/L | <0.05 |
| Zinc | mg/L | <2 |
| TOC | mg/L | <2 |
| Total coliforms/100 mL |  | 0 |
| Giardia |  | 0 |
| Cryptosporidium |  | 0 |
| Viruses |  | 0 |

* mg/L = ppm.

## HIGHLIGHTS

We have learned the following from this case study:

1. Membranes provide the ability to treat reclaimed water to a very high quality suitable for indirect potable reuse.
2. Microfiltration provides excellent pretreatment for RO.
3. Biofouling of both the MF and RO membranes is controlled through addition of a chloramine residual.
4. Reverse osmosis is capable of producing drinking water quality product water.

# Glossary

Membranes and membrane systems have their own unique, evolving terminology. This glossary provides the reader with definitions of terms used in this publication.

**Absorption**—A process by which a material (particle, molecule) is taken into a liquid or solid by physical or chemical action, but without any chemical reaction.

**Adsorption**—A process by which a material (particle, molecule) adheres to surfaces by physical action or forces without any chemical reaction.

**Angstrom**—Unit of length equaling $10^{-10}$ m ($10^{-4}$ μm, or $4 \times 10^{-9}$ in.). Its symbol is Å.

**Anion**—A negatively charged ion resulting from the disassociation of salts, acids, or bases in aqueous solution.

**Anisotrophic membranes**—Microporous membranes that vary in pore size. The surface with the smaller pore size is used as the filtering surface.

**Array**—Multiple interconnected stages in series. An assembly of cartridges in pressured membrane systems. (Also referred to as a train.)

**Backpulsing**—The reversal of permeate flow through membranes to flush trapped particles from membrane pores and cavities.

**Backwash**—The reversal of flow through a filtration medium. Often used as a cleaning operation that involves periodic reverse flow to remove foulants accumulated at the membrane surface.

**Bar**—International unit of pressure. 1 bar = 1 megadyne/sq cm; 1 bar = 100 000 Pa; 1 bar = 0.987 atm.

**Binders**—Chemicals used to hold or "bind" short fibers together in a cartridge filter.

**Brines**—Saline solutions with a concentration of dissolved solids exceeding that of seawater, typically more than 35 000 mg/L (35 000 ppm).

**Bubble point**—The amount of air pressure required to evacuate the largest pores of a fully wetted porous membrane.

**Cartridge**—A disposable filter element. Generally, a filter device designed to operate in the range of 0.1 to 100 μm ($4 \times 10^{-6}$ to $4 \times 10^{-3}$ in.).

**Cassette**—An assembly of membranes intended to be removed from an immersed system as a unit. (Also referred to as a membrane unit or rack.)

**Cation**—A positively charged ion resulting from the dissociation of salts, acids, or bases in an aqueous solution.

**Chelating agent**—A chemical reagent; typically a water-soluble organic molecule that reacts with metal ions to keep them in an aqueous solution, therefore increasing the solubility of a metal in water.

**Chemically enhanced backwash (CEB)**—A technique to clean the membrane where the fibers are soaked in chemical solution by stopping the backwash pumps for a fixed period; then the entire system is back flushed before resuming normal operation.

**Concentrate**—The portion of a feed stream that does not permeate membrane media in a cross flow filtration system, thus retaining the impurities or contaminants such as organic compounds, ions, and colloidal materials that get rejected by the membrane. (Also referred to as retentate or reject.)

**Contaminant removal**—The percentage of a contaminant removed from the feed stream by direct membrane filtration processes. Contaminant removal may be calculated for any parameter of interest (turbidity, total suspended solids, total organic carbon, etc.).

**Crossflow**—The application of water at high velocity tangential to the surface of a membrane to maintain contaminants in suspension.

**Crossflow membrane filtration**—The application of pressure and tangential flow to a semipermeable membrane surface for the separation of impurities from the water. These systems typically have one feed stream (influent) and two discharge streams (one for retentate and one for permeate).

**Dalton (Da)**—A unit of mass equal to 1/12th the mass of a carbon-12 atom [i.e., one atomic mass unit (amu)]; typically used as a unit of measure for the molecular-weight cutoff of ultrafiltration, nanofiltration, and reverse-osmosis membranes.

**Dead-end or depth filtration**—A mode of membrane filtration that has only one feed stream (influent) and one effluent stream (permeate, filtrate). Such systems do not have a crossflow type; however, a crossflow module can be operated as dead-end mode by shutting off the module outlet.

**Dialysis**—A separation process based on different diffusion rates of solutes across a permeable membrane without pressure application.

**Diffusion**—A process by which a solute moves from high concentration to low concentration across a semipermeable membrane as a result of concentration differences (gradient). Diffusion stops when the gradient no longer exists.

**Effluent**—Water that has passed through the membrane. (Also referred to as permeate, filtrate, or product water.)

**Electrodialysis**—A process in which ions are transferred through ion-selective membranes by means of an electromotive force from a less concentrated solution to a more concentrated solution.

**Element**—A term commonly used to describe an encased spiral-wound membrane module.

**Feed water**—Input stream to a membrane array or train. (Also referred to as feed stream or influent.)

**Feed stream**—Input stream to a membrane array or train. (Also referred to as feed water or influent.)

**Filter cake**—Accumulated particles on a membrane surface (or any filter surface).

**Filtrate**—Water that has passed through the membrane. (Also referred to as permeate, effluent, or product water.)

**Flat-sheet membranes**—Membrane cartridges composed of a series of flat membrane sheets and support plates. A single cassette can house many membrane cartridges that are slid into grooves for support.

**Flow equalization**—A system operating strategy where the effect of normal flow variations is dampened before the flow reaching the membrane treatment system, therefore, keeping the incoming flow rate within the acceptable hydraulic range of the membrane system. This can be achieved in several ways: by using flow storage tanks or lagoons to temporarily hold excess peak flows until the incoming flow drops and the stored wastewater can be pumped or otherwise fed back into the treatment train or by allowing the liquid level in the process tanks upstream of the membranes to rise to take advantage of the freeboard storage capacity.

**Flux**—The volume of water that passes through a membrane per unit time and per unit surface area of the membrane. Flux is measured in either liters per square meter per hour or gallons per day per square foot. Flux is affected by the water temperature; the flux is often normalized to a standard temperature of 25°C (77°F) to account for fluctuations in water viscosity. The flux in membrane bioreactor applications is also strongly affected by the mixed liquor concentration.

**Flux maintenance system**—This term is sometimes used to refer, collectively, to any equipment involved in relaxing, backwashing, or cleaning membranes to restore the system's flux rate.

**Fouling**—The buildup of impurities (such as colloidal materials) on the membrane. Fouling reduces the flux through membranes; it causes the increase of transmembrane pressure.

**Hardness**—The concentration of any multivalent cations in water. It is primarily caused by the presence of calcium and magnesium. It is measured in terms of milligrams per liter (parts per million) as calcium carbonate.

**Hollow-fiber module**—A configuration in which hollow-fiber membranes are bundled longitudinally and either encased in a pressure vessel or submerged in a basin; typically associated with microfiltration and ultrafiltration membrane processes.

**Immersed membrane systems**—The process tank(s) with interconnecting piping for feed water and drains, membrane units with air scour and filtrate connections, and frames used to support the membrane units. For pressurized systems, process tanks and frames are replaced with pressure vessels used to contain the membranes and racks or frames on which the pressure vessels are mounted. For crossflow ultrafiltration systems, piping and connections for the concentrate will also be provided.

**Influent**—Input stream to a membrane array or train. (Also referred to as feed stream or feed water.)

**Integral asymmetric membranes**—Membranes cast in one process that consist of a very thin, less than 1-μm, (less than $4 \times 10^{-5}$-in.) layer referred to as the "skin" and a thicker, up to 100-mm, (up to $4 \times 10^{-3}$-in.) porous layer that adds support and is capable of high water flux. (Also referred to as skinned membranes.)

**Integrity breach**—Whenever leaks take place that will result in the contamination or deterioration of the product (effluent quality). Under the Safe Drinking Water Act, it is defined under the Long Term 2 Enhanced Surface Water Treatment Rule (LT2ESWTR). Additionally, ASTM has developed a draft for the Standard Practice for Integrity Testing of Water Filtration Membrane Systems, D6908.

**Integrity monitoring system**—A system provided to monitor for the integrity breach of a membrane. Typically, each manufacturer may have its own system to monitor for an integrity breach.

**Isotropic membranes**—Microporous membranes that are uniform in pore size. (Also referred to as symmetric membranes.)

**Langelier Saturation Index (LSI)**—An expression to predict the precipitation of calcium carbonate under certain conditions of temperature, pH, hardness, alkalinity, and total dissolved solids.

**Lumen**—The interior of a hollow-fiber membrane.

**Macrofouling**—The buildup of impurities (such as colloidal materials) on the external membrane surface.

**Mass-transfer coefficient (MTC)**—Volume unit or mass transferred through a membrane based on a driving force.

**Membrane bioreactor (MBR)**—A combination of suspended-growth activated sludge biological treatment and membrane filtration equipment performing the critical solids/liquid separation function that is traditionally accomplished using secondary clarifiers.

**Membrane frames**—The support structure on which membrane elements are mounted.

**Membrane unit**—An assembly of membranes intended to be removed from an immersed system as a unit. (Also referred to as cassette or rack.)

**Microfouling**—The buildup of impurities (such as colloidal materials) inside the membrane's pores. (Also referred to as pore fouling.)

**Microfiltration (MF)**—A pressure-driven membrane filtration process that typically uses hollow-fiber membranes with a pore size range of approximately 0.1 to 0.2 $\mu$m ($4.0 \times 10^{-6}$ to $8.0 \times 10^{-6}$ in.).

**Micron**—A metric measurement equivalent to $10^{-6}$ m ($4.0 \times 10^{-5}$ in.). It is represented by the Greek letter $\mu$ (mu). Used to classify the pore size in a membrane.

**Microporous membranes**—Membranes cast from one material (they are homogeneous) and they can be either uniform in pore size (isotropic) or vary in pore size (anisotropic).

**Module**—A collection of membranes and housing units intended to be mounted and replaced as a unit. Sometimes in a spiral-wound membrane, the membrane element itself is called a module.

**Molecular-weight cutoff (MWCO)**—The molecular weight of the smallest material rejected by a membrane. Typically measured in Daltons. (Also referred to as nominal molecular weight cutoff.)

**Nominal molecular-weight cutoff**—The molecular weight of the smallest material rejected by a membrane. Typically measured in Daltons. (Also referred to as molecular-weight cutoff.)

**Nanofiltration (NF)**—A pressure-driven membrane separation process that uses the principles of reverse osmosis to remove dissolved contaminants from water; typically applied for membrane softening or the removal of dissolved organic contaminants.

**Osmosis**—The natural spontaneous flow of water from a less concentrated solution to a more concentrated solution across a semipermeable membrane; the flow stops when equilibrium is attained.

**Osmotic pressure**—The potential energy difference between two solutions separated by a semipermeable membrane. Osmotic pressure is a factor used in designing reverse-osmosis membrane separation equipment.

**Pascal (Pa)**—A unit that represents pressure. Pascal is the same as a Newton per square meter; thus $Pa = N/m^2$ or $psi = kPa/6.8948$.

**Permeability**—The ability of a membrane barrier to allow the passage or diffusion of a substance. It is defined as the ratio of flux to transmembrane pressure.

**Permeable**—Allowing some material to pass through.

Permeate—The water that has passed through the membrane. (Also referred to as effluent, filtrate, or product water.)

Pore—The opening in a membrane. The pore size in a membrane provides an idea of the average or smallest particle retained by a membrane.

**Pressure system frame**—The structure that supports a group of pressure vessels that are interconnected by common feed water, filtrate, and drain blocks or piping. Feed water is pumped into the system, and the residual pressure is used to convey the filtrate to filtrate storage or to a filtrate pumping station if higher pressure is required. For systems with multiple frames, the process connections from each frame are connected into larger piping manifolds.

**Product water**—The water that has passed through the membrane. (Also referred to as permeate, effluent, or filtrate.)

**Programmable logic controller (PLC)**—Controller in membrane systems that provides a variety of critical functions, including monitoring of equipment alarms and setpoints, trending of operating information such as transmembrane pressure and flow, fail-safe control and shutdown of equipment, automated control of certain operating procedures, and execution of operator-initiated or event-triggered activities.

**Rack**—An assembly of membranes intended to be removed from an immersed system as a unit. (Also referred to as membrane unit or cassette.)

**Recovery**—The ratio of the permeate flow to the feed flow; it is expressed, in general, as a percentage. Membrane bioreactor systems do not refer to recovery.

**Reject**—The portion of a feed stream that does not permeate membrane media in crossflow filtration, thus retaining impurities or contaminants such as organic compounds, ions, and colloidal materials that get rejected by the membrane. (Also referred to as concentrate or retentate.)

**Rejection**—Term used in a crossflow membrane system to express the retention of contaminants at the membrane that are larger than the pore size of the membrane.

**Retentate**—The portion of a feed stream that does not permeate membrane media in a crossflow filtration, thus retaining impurities or contaminants such as organic compounds, ions, and colloidal materials that get rejected by the membrane. (Also referred to as concentrate or reject.)

**Reverse osmosis (RO)**—A process whereby water is cleaned by forcing water through an ultrafine, semipermeable membrane that allows only water to pass though and retains contaminants; these filters are sometimes used in tertiary treatment and to pretreat water in chemical laboratories.

**Reverse-osmosis (RO) system**—Membrane system used to remove soluble ions, dissolved solids, and organic materials from high-quality tertiary effluent to polish final effluents for reuse or for groundwater recharge.

**Scaling**—The buildup of precipitated solids (salts) on a surface, such as membranes.

**Semipermeable membrane**—A membrane that allows only water to pass through while rejecting certain impurities, such as colloidal or dissolved materials.

**Silt Density Index (SDI)**—A test used to assess the treatability of a specific feed water with nanofiltration and reverse-osmosis membranes. It consists of the time required to filter 500 mL (30.5 cu in.) of water through a 0.45-$\mu$m ($1.8 \times 10^{-5}$-in.) filter. If the SDI of a feed water is too high, pretreatment before nanofiltration/reverse-osmosis will be required.

**Size exclusion**—The removal of particulate matters by a sieving mechanism.

**Skid**—A group of pressure vessels that share common valving and that can be isolated as a group for testing, cleaning, or repair.

**Skinned membranes**—Membranes cast in one process that consist of a very thin, less than 1-mm ($4.0 \times 10^{-5}$-in.), layer referred to as the "skin" and a thicker, up to 100-mm ($4.0 \times 10^{-3}$-in.), porous layer that adds support and is capable of high water flux. (Also referred to as integral asymmetric membranes.)

**Solutes**—The materials (such as chemicals) contained in a solution.

**Solvent**—The water (or any liquid) that contains dissolved matters or total dissolved solids. A solution is made up of the solvent and the solute.

**Spacers**—A meshlike material used in flat-sheet modules, such as in spirals, plate, and pleated sheet, to separate successive layers of membranes. Spacers are very important in membrane systems as they control the feed channel dimensions in membrane modules.

**Spiral-wound membrane modules (or elements)**—A membrane configuration whereby semipermeable membrane sheets are put together with a support matrix and a spacer and then wrapped around a central tube to collect the filtrate. They are primarily used in nanofiltration and reverse-osmosis membrane processes.

**Stage**—One portion of a train or array that includes membranes operating in series.

**Symmetric membranes**—Microporous membranes that are uniform in pore size (Also referred to as isotropic membranes.)

**System arrays**—The trains or arrays needed to produce the design flow of a plant.

**Thin-film composite (TFC) membranes**—Membranes made by bonding a thin cellulose acetate, polyamide, or other acetate layer, typically 0.15 to 0.25 μm thick ($6.0 \times 10^{-6}$ to $9.8 \times 10^{-6}$ in.), to a thicker porous substrate, which provides structural stability.

**Train**—Multiple interconnected stages in series. An assembly of cartridges in pressured systems. (Also referred to as array.)

**Transmembrane pressure (TMP)**—The difference between the average feed/concentrate pressure and the permeate pressure; the driving force, or hydraulic head loss, associated with any given flux. The TMP of the membrane system is an overall indication of the feed pressure requirement; it is used, with the flux, to assess membrane fouling and the need for chemical cleaning. In a crossflow membrane system, TMP is measured as the average of the inlet and outlet pressures, minus the permeate backpressure.

**Tubular systems**—Membrane systems in which the membranes are cast on the inside of a support tube and then placed into a pressure vessel. The feed water is pumped through the feed tube and the product water is collected on the outside of the tubes, while the concentrate continues to flow through the feed tube.

**Ultrafiltration (UF)**—A pressure-driven membrane filtration process that typically uses hollow-fiber membranes with a pore size range of approximately 0.01 to 0.05 μm ($4.0 \times 10^{-7}$ to $2.0 \times 10^{-6}$ in.).

**Vessel**—A pressurized tube that contains several membrane elements in series.

# APPENDIX I

# Related Equations

## Flux

Flux is defined as the permeate flow divided by the total membrane surface area, as shown in the formula below, and is often presented in units of liters per square meters of membrane surface area per hour (gallons per day per square foot). Because the flux is greatly affected by the water temperature, the flux is often normalized to a standard temperature of 25° C (77° F) to account for fluctuations in water viscosity:

$$J = \frac{Q_P}{A_{System}}$$

Where
- $J$ = flux (L/m²·h [gpd/sq ft]),
- $Q_P$ = permeate flow (L/h [gpd]), and
- $A_{System}$ = surface area of the membrane system (m² [sq ft]).

Flux can also be calculated as follows:

$$J = A_P(P_T - P_O)$$

Where
- $A_P$ = the membrane permeability coefficient, which is the reciprocal of resistance to flow (also referred to by the letter $K$ in other references, or mass-transfer coefficient [MTC] as defined below) (dimensionless),

$P_T$ = the transmembrane pressure (TMP) (kPa [psi]), and
$P_O$ = the osmotic pressure of the feed solution (kPa [psi]).

The above equation clearly shows that for water (feed solution) to flow through the membrane, the TMP must be greater than the osmotic pressure of the solution to provide a positive driving force.

The effect of temperature on flux can be evaluated as follows:

$$J_T = J_{25} \times 1.03^{(T-25)}$$

Where

$J_T$ = the flux at temperature $T$ (°C) and
$J_{25}$ = the flux at temperature 25°C (77°F).

The 25° C (77° F) reference temperature is used in nanofiltration (NF)/reverse osmosis (RO). For MF/UF, 20°C (68°F) is used for correcting the permeate flux to a reference temperature. The water flux typically increases by 3% for each degree temperature increase.

Therefore, to be able to evaluate changes in system in performance over time, all data must be "normalized" to a constant temperature. Figure AI.1 below shows an example of a normalized plot of MTCs for a pressurized MF system treating unchlorinated secondary effluent. For low-pressure membrane processes, common practice is to normalize flux data to 20° C (68° F) using one of the following equations. The expressions within brackets are correlations of viscosity with temperature:

$$J_{20} = J_T \cdot \text{TCF} \rightarrow \text{TCF} = \frac{\mu_T}{\mu_{20}}$$

$$J_{20} = J_T \cdot \left[ 1.784 - (0.0575 \cdot T) + (0.001 \cdot T^2) - (10^{-5} \cdot T^3) \right]$$

or

$$J_{20} = J_T \cdot e^{[-0.032 \cdot (T-20)]}$$

Where

$J_{20}$ = normalized flux at 20°C (L/m²·h [gpd/sq ft]),
$J_T$ = actual flux at temperature T (L/m²·h [gpd/sq ft]),

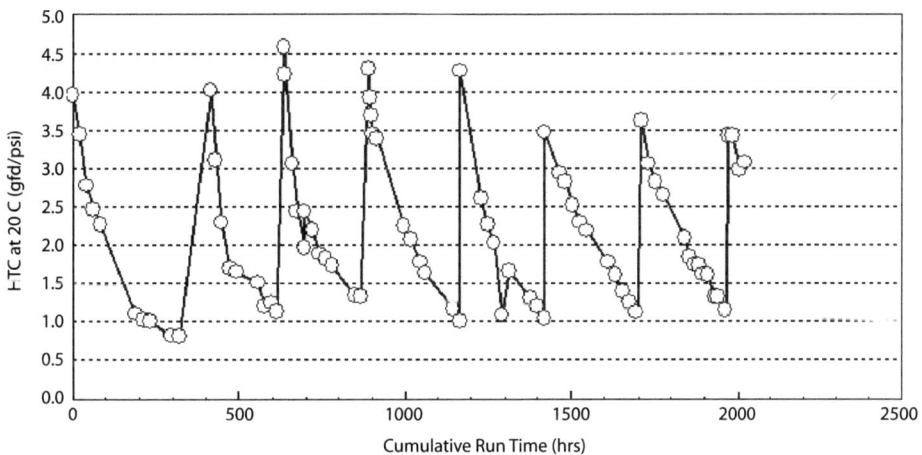

**FIGURE AI.1—EXAMPLE NORMALIZED MASS-TRANSFER COEFFICIENT PLOT FOR A PRESSURE MF SYSTEM** (gpd/sq ft/psi × 0.246 = $\frac{L/m^2 \cdot h}{kPa}$) **(COURTESY OF WATER ENVIRONMENT RESEARCH FOUNDATION).**

$T$ = water temperature (°C),
TCF = temperature correction factor,
$\mu_{20}$ = viscosity of water at 20°C (cP [Pa·s]) = 1.0, and
$\mu_T$ = viscosity of water at temperature $T$, °C (cP [Pa·s]).

## Total Surface Area

The total surface area represents the total membrane surface area available for treatment in a membrane system. This may be calculated by multiplying the surface area contained within each element by the number of elements contained in the membrane system, as shown by:

$$A_{System} = A_{Element} \times N_{Element}$$

Where

$A_{System}$ = total surface area (m² [sq ft]),
$A_{Element}$ = surface area of each element (m² [sq ft]), and
$N_{Element}$ = number of elements in membrane system.

## Transmembrane Pressure

The TMP is defined as the difference between the average feed/concentrate pressure and the permeate pressure, as shown below. It is effectively the driving force for flux. The TMP of the membrane system is an overall indication of the feed pressure requirement; it is used, with the flux, to assess membrane fouling.

For crossflow mode of operation:

$$\text{TMP} = \left(\frac{P_F + P_C}{2}\right) - P_P$$

Where
- TMP = transmembrane pressure (kPa [psi]),
- $P_F$ = feed pressure (kPa [psi]),
- $P_C$ = concentrate pressure (kPa [psi]), and
- $P_P$ = permeate pressure (kPa [psi]).

For the direct-feed mode of operation:

$$\text{TMP} = P_F - P_P$$

It should be noted that, with MF/UF, the feed pressure and concentrate pressure are equal.

## Net Driving Pressure

Net driving pressure (NDP) is the pressure available to drive the feed water through the membrane minus the permeate and osmotic backpressure:

$$\text{NDP} = P_F - \frac{1}{2}(P_F - P_C) - P_P - \Delta\pi$$

Where
- $P_F$ = feed pressure (kPa [psi]),
- $P_C$ = concentrate pressure (kPa [psi]),
- $P_P$ = permeate or filtrate pressure (kPa [psi]), and
- $\Delta\pi$ = change in osmotic pressure (kPa [psi]).

# Recovery

The percentage of feed that is converted to permeate is called the recovery (water or liquid) of the membrane system and is calculated by the formula below (see Figure AI.2). Recovery is critical in NF and RO, as not all of the liquid will go through the membrane surface. In MF and UF, the liquid stream applied to the membrane surface will go through. Generally, it is desired to operate at a high recovery, as it minimizes the waste stream. However, operating at elevated recoveries may result in increased fouling rates and cleaning frequencies. Manufacturers should be consulted before altering the operating recovery of the membrane system.

$$R = \frac{Q_P}{Q_F} \times 100\%$$

Where

- $R$ = recovery (%),
- $Q_P$ = permeate flow (L/s [gpm]), and
- $Q_F$ = feed flow (L/s [gpm]).

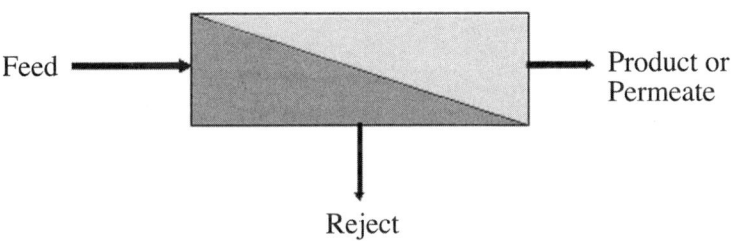

FIGURE AI.2—SCHEMATIC DEPICTING PERCENT RECOVERY.

## Contaminant Removal (or Rejection)

Contaminant removal is defined as the percentage of a contaminant removed from the feed stream by the membrane and may be calculated by the formula shown below. Contaminant removal may be calculated for any parameter of interest (turbidity, total suspended solids, total organic carbon, etc.); however, consistent units must be maintained throughout the calculation:

$$R_{Cont} = \frac{C_F - C_P}{C_F} \times 100\%$$

Where

$R_{Cont}$ = contaminant removal (%),
$C_F$ = feed contaminant concentration (e.g., mg/L), and
$C_P$ = permeate contaminant concentration (e.g., mg/L).

## Backwash Volume

Microfiltration (MF) and ultrafiltration (UF) backwash volumes can be estimated with the following equation:

$$V = \text{No. of backwashes per day} \times \text{Volume per backwash}$$
$$= \frac{1440 BW_V}{BW_{int}}$$

Where

$BW_V$ = volume of water used per backwash (L) and
$BW_{int}$ = backwash interval in minutes (minutes).

A mass balance around the membrane process can be used to estimate the concentration of feed water constituents in the backwash water according to the following equation:

$$C_{BW} = \frac{C_f}{1-r}$$

Where

$C_f$ = concentration of constituent in the feed water and
$r$ = system recovery.

## Langelier Saturation Index

The most common method for determining the solubility of calcium carbonate ($CaCO_3$) in water is the Langelier saturation index (LSI). Waters that are negative on this index indicate an undersaturated condition with respect to $CaCO_3$ (tendency to dissolve $CaCO_3$), whereas waters that are positive indicate an oversaturated condition (tendency to precipitate $CaCO_3$). The LSI equation is as follows:

$$LSI = pH - pH_s$$
$$pH_s = (9.3 + A + B) - (C + D)$$

Where

$pHs$ = saturation pH for $CaCO_3$,
$A = \dfrac{(\log_{10}[\text{Total dissolved solids}] - 1)}{10}$
$B = -13.12 \times \log_{10}(°C + 273) + 34.55$,
$C = \log_{10}[Ca^{+2}\text{ as }CaCO_3] - 0.4$, and
$D = \log_{10}[\text{Alkalinity as }CaCO_3]$.

## Stiff and Davis Scaling Index

Other equations used for determining the solubility of calcium carbonate include the Stiff and Davis scaling index (SDSI). The SDSI equation is as follows:

$$SDSI = pH - p_{Ca} - p_{Alk} - K$$

Where

$p_{Ca}$ = negative logarithm of the calcium molarity, as in LSI;

$p_{Alk}$ = negative logarithm of the alkalinity in equivalents per liter, as in LSI; and

$K$ = different empirical constant from LSI to account for temperature and ionic strength.

## Silt Density Index

The tendency for the water feed to foul a membrane can be evaluated with a filterability test called the silt density index (SDI). The SDI is primarily applicable in NF/RO. The test is described in the ASTM standard no. D4185. The test is very simple and consists of a 0.45-μm ($1.8 \times 10^{-5}$-in.) cellulose acetate membrane in a dead-end filtration cell. The test is conducted for 15 minutes, labeled the total test time, $T_t$ or $T_{15}$. The time, in minutes, needed to collect the initial 500 mL (30.5 cu in.) of filtrate is recorded as $T_i$. The time needed to collect another 500 mL (30.5 cu in.) of filtrate after the filter has been online for 15 minutes is recorded as $T_f$. Standard conditions for the SDI determination call for a 47-mm-diam (1.85-in.-diam) filter and an applied pressure of 206.8 kPa (30 psi or 2 bar) and a total test time of 15 minutes. The SDI is calculated according to

$$SDI = \frac{100\left(1 - \frac{T_i}{T_f}\right)}{T_t}$$

Where

$T_i$ = time in minutes needed to collect the initial 500 mL (30.5 cu in.),

$T_f$ = time needed to collect 500 mL (30.5 cu in.) after being online for 15 minutes, and

$T_t$ = total time of the test, 15 minutes.

For successful operation of hollow-fiber and spiral-wound RO membranes, a SDI of 2 to 3 is desirable in desalting membranes, with 3 to 5 as an upper limit. For best performance, membrane manufacturers will recommend the SDI limit. Exceeding the SDI limit will require pretreatment of the feed to the membrane and may require some operational changes, such as reduced flux rate. With lower SDI, pretreatment may be advisable, as it reduces the cleaning cycle.

# Solubility Product

The solubility product for a given salt can be found using the following equation:

$$K_{sp} = [\text{Cation}]^{\#} \times [\text{Anion}]^{\#}$$

Where

$K_{sp}$ = a value for the solubility product,
[Cation] = the cation concentration,
[Anion] = the anion concentration, and
\# = the quantity of the particular ion present within the salt molecule.

# APPENDIX II

# References and Resources

Abi-Samra, B. C.; Cook, P.; Lange, P.; Zepeda, J. (2001) The Use of Microfiltration and Reverse Osmosis to Treat Secondary Wastewater Treatment Plant Effluent and Provide High Quality Reclaimed Water. *Proceedings of the American Water Works Association Annual Conference* [CD-ROM]; Washington, D.C., June 17–21; American Water Works Association: Denver, Colorado.

Alexander, K. (2003) Emergent Contaminant Removal using Reverse Osmosis for Indirect Potable Reuse. Paper presented at the International Desalination Association World Congress; Paradise Island, Bahamas, Sep 28–Oct 3; International Desalination Association: Topsfield, Massachusetts.

Allgeier, S.; Alspach, B.; Vickers, J. (2003) *Draft Membrane Filtration Guidance Manual.* U.S. Environmental Protection Agency: Cincinnati, Ohio.

American Public Health Association; American Water Works Association; Water Environment Federation (19989) *Standard Methods for the Examination of Water and Wastewater*, 20th ed.; American Public Health Association: Washington, D.C.

American Water Works Association (1990) *Water Quality and Treatment*, 4th ed.; American Water Works Association: Denver, Colorado.

American Water Works Association (1999) *Manual of Water Supply Practices, Reverse Osmosis and Nanofiltration*; Manual M46; American Water Works Association: Denver, Colorado.

American Water Works Association (2003) Residuals Management for Low-Pressure Membranes. Committee Report. *J. Am. Water Works Assoc.*, **95** (6).

American Water Works Association (2004) Current Perspectives on Residuals Management for Desalting Membranes. Committee Report. *J. Am. Water Works Assoc.*, **96** (12).

American Water Works Association Research Foundation; Lyonnaise des Eaux; Water Research Commission of South Africa, Eds. (1996) *Water Treatment Membrane Processes.* McGraw-Hill: New York.

Amjad, Z., Zibrida, J.; Zuhl, R. (1997) A New Antifoulant for Controlling Silica Fouling in Reverse Osmosis Systems. *Proceedings of the International Desalination Association World Congress on Desalination and Water Reuse* Vol. III; 459–480; Madrid, Spain, Oct 6–9; International Desalination Association: Topsfield, Massachusetts.

Amjad, Z., Zibrida, J.; Zuhl, R. W. (1999) Reverse Osmosis Technology: Fundamentals and Water Applications. *The Analyst*, Fall 1999, Association of Water Technologies: McLean, Virginia.

Anonymous (2002) *Singapore Water Reclamation Study—Expert Panel Review and Findings.* http://www.pub.gov.sg/NEWater_files/download/review.PDF.

Asano, T., Ed. (1998) *Wastewater Reclamation and Reuse.* Technomic Publishing: Lancaster, Pennsylvania.

ASTM International (2002) *An American National Standard Practice for Integrity Testing of Water Filtration Membrane Systems*; D 6908 03; ASTM International: West Conshohocken, Pennsylvania.

Beaton, N. C.; Cooper A. R. (1980) Ultrafiltration Membranes and Applications. *Polym. Sci. Technol.*, **13,** 373–404.

Bourgeous, K.; Darby, J.; Tchobanoglous, G. (1998) Membrane Filtration for Water Reuse: An Evaluation of Ultrafiltration As a Secondary and Tertiary Treatment Process. *Proceedings of the Microfiltration II Conference*; San Diego, California, Nov 12–13; National Water Research Institute: Fountain Valley, California; 173–187.

Byrne, W. (1995) Reverse Osmosis, *A Practical Guide for Industrial Users.* Tall Oaks Publishing: Littleton, Colorado.

Chalmers, R. B.; Leslie, G.; Sudak, D.; Alexander, K. (2000) Selection of a Microfiltration Process for the Groundwater Replenishment System, the Largest Advanced Recycled Water Treatment Plant in the

World. *Proceedings of the American Water Works Association Annual Conference* [CD-ROM]; Denver, Colorado, June 11–15; American Water Works Association: Denver, Colorado.

Cheryan, M. (1998) *Ultrafiltration and Microfiltration Handbook*. Technomic Publishing: Lancaster, Pennsylvania.

Conlon, W. J.; McClellan, S. A. (1989) Membrane Softening, A Water Treatment Process That Has Come of Age. *J. Am. Water Works Assoc.*, **81** (11), 47–51.

Consept: Consultant in Separation Technology, Drinking Water Inspectorate (2001) Review of the Adequacy of Existing Proposals for Membrane Integrity Monitoring; Final Report; 42/2/159; Consept: United Kingdom.

Crites, R.; Tchobanoglous, G. (1998) *Small and Decentralized Wastewater Management Systems*, 1st ed. McGraw-Hill: New York.

Dawes, T. M. (2002) Selecting Microfiltration Equipment for the Orange County Groundwater Replenishment System. *Proceedings of the Microfiltration III Conference—A Technology Comes of Age*; Costa Mesa, California, May 5–7; National Water Research Institute: Fountain Valley, California; 93–95.

Deshmukh, S. S.; Patel, M. V.; Everest, W. R.; Daugherty, J. L. (2003) Emerging Contaminant Removal using RO and UV for the Groundwater Replenishment System. *Proceedings of the American Water Works Association Membrane Technology Conference*; Atlanta, Georgia, Mar 2–5; American Water Works Association: Denver, Colorado.

Drewes, J. E.; Herberer, T.; Reddersen, K. (2002) Fate of Pharmaceuticals during Indirect Potable Reuse. *Water Sci. Technol.*, **46**, 73–80.

Drewes, J. E.; Xu, P.; Bellona, C.; Amy, G.; Kim, T.; Adam, M.; Heberer, T. (2004) Rejection of Emerging Organic Micropollutants in Nanofiltration/Reverse Osmosis Membrane Applications. *Proceedings of the 77th Annual Water Environment Federation Technical Exhibition and Conference* [CD-ROM]; New Orleans, Louisiana, Oct 2–6; Water Environment Federation: Alexandria, Virginia.

Duranceau, S. J. (1994) Mathematics for Membrane Processes. *Fla. Water Resource. J.*, **46** (9), 26–28.

Fleischer, E. J.; Broderick, T. A.; Daigger, G. T.; Lozier, J. C.; Wollmann, A. M.; Fonseca, A. D. (2001) Evaluating the Next Generation of Water Reclamation Processes. *Proceedings of the 74th Annual*

*Water Environment Federation Technical Exposition and Conference* [CD-ROM]; Atlanta, Georgia, Oct 13–17; Water Environment Federation: Alexandria, Virginia.

Freeman, S.; Harvey, G. (1999) As Easy as 1, 2, 3, .... *Ind. Wastewater*, **Sept/Oct,** pp 30–34.

Freeman, S.; Harvey, G., Adams, L. (2005) Eight Years and Two Billion Gallons Later: An Update on a Triple Membrane Plant. *Proceedings of the American Water Works Association Membrane Technology Conference*; Phoenix, Arizona, March.

Freeman, S. D. N.; Craig, M.; Hemken, B. E. (1996) Wastewater Reclamation Using the Microfiltration and Reverse Osmosis/Nanofiltration Process at Scottsdale, Arizona. *Proceedings of the Joint American Water Works Association/ Water Environment Federation Water Reuse Conference*; San Diego, California, May; American Water Works Association: Denver, Colorado.

Geselbracht, J. (1994) A Comparison of Equipment for the Microfiltration of Secondary Effluent. *Proceedings of the Microfiltration for Water Treatment Symposium*; Irvine, California, Aug 25–26; National Water Research Institute/Orange County Water District: Fountain Valley, California; 65–76.

Geselbracht, J. (1998) Membrane Treatment for Water Reuse in the Livermore—Amador Valley. *Proceedings of the Microfiltration II Conference*; San Diego, California, Nov 12–13; National Water Research Institute: Fountain Valley, California; 55–63.

Huber, M. M.; Canonica, S.; Park, G. Y.; Gunten, U. V. (2002) Oxidation of Pharmaceuticals during Ozonation and Advanced Oxidation Processes. *Environ. Sci. Technol.*, **37**, 1016–1024.

Hydranautics (2002) *Technical Service Bulletin 139.01—Data Logging, Normalization, and Performance Analysis for HYDRAcap Systems.* Hydranautics: Oceanside, California.

Indeck Water Treatment Systems (2002) *Operations and Maintenance Manual for the Cogentrix-Caledonia Generating Facility*; Indeck Water Treatment Systems, Wheeling, Illinois.

International Union of Pure and Applied Chemistry (1997) *Compendium of Chemical Terminology*, 2nd ed.; Blackwell Science: Cambridge, Massachusetts.

Kavanaugh, M. C. (2003) Unregulated and Emerging Chemical Contaminants: Technical and Institutional Challenges. *Proceedings of the 76th Annual Water Environment Federation Technical Exhibition and Conference* [CD-ROM]; Los Angeles, California, Oct 11–15; Water Environment Federation: Alexandria, Virginia.

Kimura, K.; Amy, G.; Drewes, J. E.; Heberer, T.; Kim, T. K.; Watanabe, Y. (2003) Rejection or Organic Micropollutants (DBPs, EDCs, PACs) by NF/RO Membranes. *J. Membrane Sci.*, **227**, 113–121.

Kinslow, J. K.; Manning-Hudkins, J. (2004) The Evolution of Pre-Treatment Chemicals in Membrane Process: An Analysis of Innovation in the Pre-Treatment Chemical Industry. *Fla. Water Resour. J.*, **56** (11), 25–28.

Langelier, W. F. (1936) The Analytical Control of Anti-Corrosion Water Treatment, *J. Am. Water Works Assoc.*, **28**, 1500–1521.

Law, I.; Leslie, G.; Poon, J. (2002) Water Reclamation Using the Dual-Membrane System—The Singapore Experience. Graham, N. J. D. and Lai, P., Eds.; *Proceedings of the International Conference on Wastewater Management and Technologies for Highly Urbanized Coastal Cities 2002* [CD-ROM]; Causeway Bay, Hong Kong, June 10–12; The Hong Kong Polytechnic University: Hong Kong.

Leslie, G. L.; Dawes, T. M.; Snow, T. S. (2000) Developing Indirect Potable Reuse to Increase Water Supply, Improve Water Quality and Manage Wastewater Discharge in Orange County California. *Proceedings of the 73rd Annual Water Environment Federation Technical Exposition and Conference* [CD-ROM]; Anaheim, California, Oct 14–18; Water Environment Federation: Alexandria, Virginia.

Leslie, G. L.; Patel, M. V.; Norman, J. E.; Dunivin, W. R.; Martin, E. M.; Sudak, R. G. (1999) Microporous Membrane Pretreatment Options for Reverse Osmosis in Municipal Wastewater Reclamation Applications. *Proceedings of the American Water Works Association Membrane Technology Conference* [CD-ROM]; Long Beach, California, Feb 28–Mar 3; American Water Works Association: Denver, Colorado.

Lozier, J. (2000) Two Approaches to Indirect Potable Reuse Using Membrane Technology. *Water Sci. Technol.*, **41** (10–11), 149–156.

Lozier, J.; McKim, T. (1994) Microfiltration for Water Reuse and Reverse Osmosis Pretreatment. *Proceedings of the Microfiltration for Water*

*Treatment Symposium*; Irvine, California, Aug 25–26; National Water Research Institute/Orange County Water District: Fountain Valley, California; 87–89.

Lozier, J. C. (1998) *Wastewater Reclamation Pilot Study: City of McAllen, Texas*; CH2M Hill, Inc., Report No. 26; U.S. Department of the Interior, Bureau of Reclamation: Washington, D.C.

Malcolm Pirnie, Inc.; CH2M Hill, Inc.; Separation Processes, Inc. (2001) *Low-Pressure Membrane Filtration For Pathogen Removal: Application, Implementation and Regulatory Issues*. EPA-815/C-01-001; U.S. Environmental Protection Agency: Cincinnati, Ohio.

Malcolm Pirnie, Inc.; Separation Processes, Inc.; The Cadmus Group, Inc. (2003) *Membrane Filtration Guidance Manual*; Proposal Draft; EPA-815/D-03-008; U.S. Environmental Protection Agency: Cincinnati, Ohio.

Manning-Hudkins, J.; Schmidt, H. E. (2003) Membrane Treatment Processes and Their Ability to Address Emergent Pollutants of Concern. Paper presented at the South Florida Water Management District (SFWMD) Advanced Water Treatment Technologies and Emergent Pollutants of Concern Workshop, West Palm Beach, Florida, Oct 17.

Metcalf and Eddy, Inc. (2003) *Wastewater Engineering: Treatment and Reuse*, 4th ed.; McGraw-Hill: New York.

Montgomery Watson (1994) *San Diego County Water Authority Water Repurification Feasibility Study Final Report*. Montgomery Watson: Broomfield, Colorado.

Mulder, M. (1996) *Basic Principles of Membrane Technology*. Kluwer Academic Publishers: Dordrecht, The Netherlands.

Nemeth, J. E. (1997) Scale Inhibitors: Application, Developments and Trends. *Proceedings of the International Desalination Association (IDA) World Congress on Desalination and Water Reuse*; Madrid, Spain, Oct 6–9; International Desalination Association: Topsfield, Massachusetts.

Pall Corporation (2004) *Microza® Microfiltration System Installation, Operation & Maintenance Manual*. Pall Corporation: Cortland, New York.

Reardon, R. D.; DiGiano, F. A.; Aitken, M. D.; Paranjape, S. V.; Foussereau, X. J.; Kim, J. H.; Chang, S.-C.; Crammer, R. (2005) *Membrane*

*Treatment of Secondary Effluent for Subsequent Use Project No. 01-CTS-6*; Phase 1 Report; Water Environment Research Foundation: Alexandria, Virginia.

Ryznar, J. W. (1944) A New Index for Determining Amount of Calcium Carbonate Scale Formed by Water. *J. Am. Water Works Assoc.*, **36**, 472–478.

Schimmoller, L.; McEwen, B. (2001) Ferric Chloride Pretreatment Improves Membrane Performance: Pilot Results for Denver's New 30 MGD Water Reclamation Plant. *Proceedings of the Water Reuse Association Symposium XVI* [CD-ROM]; San Diego, California; Water Reuse Association: Alexandria, Virginia.

Shoenberger, P. E. (2002) West Basin Municipal Water District and the Role of Microfiltration. *Proceedings of the Microfiltration III Conference—A Technology Comes of Age*; Costa Mesa, California, May 5–7; National Water Research Institute: Fountain Valley, California; 99–101.

Snoeyink, V. L.; Jenkins, D. (1980) *Water Chemistry*. Wiley & Sons: New York.

Stephenson, T.; Brindle, K.; Judd, S.; Jefferson, B. (2000) *Membrane Bioreactors for Wastewater Treatment*. IWA Publishing: London, United Kingdom.

Sudak, R. G.; Dunivin, W. R. (1994) Pilot Plant Testing of Direct Filtration, Microfiltration and Reverse Osmosis at Water Factory 21. *Proceedings of the Microfiltration for Water Treatment Symposium*; Irvine, California, Aug 25–26; National Water Research Institute/Orange County Water District: Fountain Valley, California; 97–98.

Taniguchi, Y. (1994) State of the Art of Microfiltration Technology for Drinking and Water and Wastewater Treatment in Japan. *Proceedings of the Microfiltration for Water Treatment Symposium*; Irvine, California, Aug 25–26; National Water Research Institute/Orange County Water District: Fountain Valley, California; 13–24.

Tchobanoglous, G. (2002) Membrane Technology Overview. *Proceedings of the California Water Environment Association's Membrane Technology One-Day Specialty Conference*; California Water Environment Association: Oakland, California.

The Drinking Water Inspectorate, Department for Environment, Food and Rural Affairs (2001) *Review of the Adequacy of Existing*

Proposals for Membrane Integrity Monitoring; Final Report; 42/2/159; Consept: United Kingdom.

Trivedi, H. K. (2004) Flat-Plate Microfiltration Membrane Bioreactor Designed for Ultimate Nutrient Removal (UNR™). *Proceedings of the 77th Annual Water Environment Federation Technical Exhibition and Conference* [CD-ROM]; New Orleans, Louisiana, Oct 2–6; Water Environment Federation: Alexandria, Virginia.

U.S. Environmental Protection Agency (1993) *Manual: Nitrogen Control*; EPA-625/R-93-010; U.S. Environmental Protection Agency: Washington, D.C.

USFilter (2002) *Installation, Operation, and Maintenance Manual for 6M10C Unit*. USFilter: Warrendale, Pennsylvania.

Vernon, W. (2003) The Application of Membrane Systems in Scottsdale, Arizona. *Proceedings of the American Water Works Association Membrane Technology Conference*; Atlanta, Georgia; American Water Works Association: Denver, Colorado.

Vernon, W. (2005) Scottsdale, Arizona—A Microcosm of Communities Striving to Meet Long Term Supply and Quality Issues in a Desert Environment *Proceedings of the American Water Works Association 2005 Membrane Technology Conference*, Phoenix, Arizona, March; American Water Works Association: Denver, Colorado.

Vernon, W.; Alexander, K. L. (2000) Scottsdale Water Campus: Reuse Solutions Using Microfiltration and Reverse Osmosis. *Proceedings of the American Water Works Association Annual Conference* [CD-ROM]; Denver, Colorado, June 11–15; American Water Works Association: Denver, Colorado.

Vernon, W.; Clune, J.; Nunez, A. (2001) Applying an Integrated Membrane System to Meet Water Quality Goals For the Indirect Potable Reuse of Tertiary Effluent. *Proceedings of the American Water Works Association 2001 Membrane Technology Conference*, San Antonio, Texas, March; American Water Works Association: Denver, Colorado.

Vickers, J. C. (1998) Application and Design of Low Pressure Membrane Systems Current Status and Future Trends. *Proceedings of the Microfiltration II Conference*; San Diego, California, Nov 12–13; National Water Research Institute: Fountain Valley, California; 11–15.

Water Environment Federation (1992) *Design of Municipal Wastewater Treatment Plants*, 3rd ed.; Manual of Practice No. 8; Water Environment Federation: Alexandria, Virginia.

Water Environment Federation (1996) *Operation of Municipal Wastewater Treatment Plants*, 5th ed.; Manual of Practice No. 11; Water Environment Federation: Alexandria, Virginia.

Water Environment Research Foundation (2001) *Membrane Bioreactors: Feasibility and Use in Water Reclamation.* Final Report Project #98-CTS-5; Water Environment Research Foundation: Alexandria, Virginia.

Water Environment Research Foundation (2005) *Membrane Treatment of Secondary Effluent for Subsequent Use—Progress Report*; Project 01-CTS-6; Water Environment Research Foundation: Alexandria, Virginia.

Won, W.; Shields, P. (1999) Comparative Life Cycle Costs for Operation of Full Scale Conventional Pretreatment/RO and MF/RO Systems. *Proceedings of the American Water Works Association Membrane Technology Conference* [CD-ROM]; Long Beach, California, Feb 28–Mar 3; American Water Works Association: Denver, Colorado.

ZENON Environmental, Inc. (2003) *Operation and Maintenance Manual Ultrafiltration Wastewater Treatment System for Rubes Creek Wastewater Treatment Plant, City of Woodstock, Georgia.* ZENON Environmental, Inc.: Oakville, Ontario, Canada.

# Index

## A

absorption, 229
acids, humic, 20
acoustic sensors, low-pressure
  memranes, 142
activated sludge
  aeration system, 58
  monitoring, 97–99
  quality, 72
active management areas, 222, 223
adsorption, 229
advanced wastewater treatment, 2
aeration, 38, 42, 95
  activated sludge, 58
  coarse-bubble, 77, 205
  cyclic, 86
  fine bubble, 189
  diffusers, 77, 78
  jet, 77, 86
  membrane bioreactor, 77, 208
  reverse osmosis, 160
  tanks, 12
aerobic treatment, 11
agricultural reuse, 8
air compressors, 87
air scouring, 17, 21, 40, 42, 50, 95
  blowers, 85, 86

membrane bioreactor, 50, 51
optimization, 96, 97
troubleshooting, 110
air separator, 82
alarms, 95
  equipment and instrumentation, 101
  high and low tank levels, 95
  high flow, 95
  mixed liquor recycle flow, 95
  pH, 168, 169
  posttreatment, 169
  pretreatment, 169
  reverse osmosis, 169
  scour airflow, 95
  transmembrane pressure, 48, 49, 133, 134
  turbidimeters, 88, 89
  turbidity, 95
algae, 20, 39, 130
alkalinity, membrane bioreactor, 63, 64, 91, 92
AMA, 222, 223
anaerobic treatment, 11
anaerobic zone, 99, 188
analysis, source water, 153
anion, 229
anisotrophic membranes, 229

anisotropic pore size, 24
anoxic zone, 188
anoxic zone, sampling, 99
antifoulants, silica, 167
antiscalants, 216, 217
antiscalants, reverse osmosis, 157, 158
applications
    groundwater recharge, 216
    membrane, 10–12
aqueous salts, 20
Arizona Groundwater Management Act, 222
array, 21, 229
    two and three staged, 162
    two-pass, 163
asymmetric membranes, 24, 26
audible test, low-pressure membranes, 142
automated systems
    filtrate flow control, 132, 133
    operational considerations, 131–134
automatic fine screening, 37, 38

## B

backpulse, 93
    permeability, 100
    pumps, 83
    tanks, 83, 84
backpulsing, 230
backwash, 93, 230
    microfiltration, 246, 247
    procedure, 36
    ultrafiltration, 246, 247
    volume, 246, 247
    water, 125
backwash system, 42, 43, 49, 50, 133
    backwash flux, 49
    chemical addition, 49
    chemically enhanced backwashes, 43
    cross-flow velocity, 49
    programmable logic controllers, 48
    transmembrane pressure, 49
backwashing, 41–43, 141
    CEB, 50
    chemically enhanced, 50
    immersed membranes, 50, 51
    inside-out flow, 50
backwashing frequency, 136
    optimization, 136
    transmembrane pressure, 136
bacteria, 18, 20, 150
barium sulfate, 166
bicarbonate, 150, 151
binders, 230
biochemical oxygen demand
    membrane bioreactor, 192, 198, 203, 208
    microfiltration and ultrafiltration, 117, 118
biofilm, reverse osmosis, 170
biofouling, chloramines, 227
biological growth, reverse osmosis, 170
biological nutrient removal, 68, 70, 71
biological oxygen demand, 150, 213
biological process operation
    dissolved oxygen, 91
    membrane bioreactor, 89–92
biological treatment
    applications, 67, 68
    membrane bioreactor, 67–73
bioreactors, 12
biosolids, membrane bioreactor, 75
blowers, 76, 77
    air scour, 85, 86
    optimization, 96, 97
    troubleshooting, 110

bubble point, 230
bubble point test, 46
bubble-point theory, 45
    diffusive airflow test, 45, 46
    pressure decay test, 45, 46

## C

calcium carbonate
    reverse osmosis, 165
    solubility, 247, 248
calcium sulfate, 166
calibration, instruments, 108, 109, 144, 180, 181
capacity management, 96
capacity management, reverse osmosis, 170, 171
cartridge filters, 39, 153, 154, 178, 230
cartridge filters, reverse osmosis, 210–212, 224, 225
cartridges, 21, 28
CAS, 13
case histories, overview, 185
cassette, 22, 29, 30, 230
cation, 230
CEB, 43, 50, 52, 93, 102, 230
cellulose, 27
ceramic membrane, 26
challenge tests, low-pressure membranes, 142
Chandler Arizona Reverse Osmosis Wastewater Reclamation Plant, 214–221
    plant data and design, 218–221
chemical addition, programmable logic controllers, 48
chemical cleaning solutions, low-pressure membranes, 126
chemical cleaning, 17, 40, 51, 52
    chemicals used, 126
    frequency, 126, 137

low-pressure membranes, 141
microfiltration, 211, 224
reverse osmosis, 169, 170, 177, 178, 211, 220, 221, 225
solutions used, 52
systems, 44, 159
typical chemicals, 124
chemical feed systems
    coagulants, 74
    low-pressure membranes, 124
    membrane bioreactor, 74
    sodium hydroxide, 74
chemical feed, setpoints, 164
chemical injection systems, operational considerations, 131
chemical metering
    monitoring frequency, 138
    station, 85
chemically enhanced backpulse, 93, 102
chemically enhanced backwash, 43, 50, 52, 230
chemicals
    cleaning, 84, 85
    operational considerations, 131
    optimization, 97
    safety, 152
    storage, 124
    typical for low-pressure membranes, 124
chemistry, reverse osmosis, 165
chloramination, 153
chloramines, 154, 155, 211
chloramines, biofouling control, 214, 227
chlorination, 154–156
chlorination, low-pressure membranes, 122
chlorine, 11, 27
    free, 155
    tolerance, 35
CIP, 43, 51, 52

backpulse pumps, 83, 84
    membrane bioreactor, 74
citric acid, 85, 211
    cleaning, 102
    recovery cleaning, 105
clarifiers, secondary, 12
Clean Water Act, 3, 117
cleaning
    agents, reverse osmosis, 177
    chemicals, 17, 42, 124, 177, 178
    diffusers, 106
    factors that influence recovery cleaning time, 103, 104
    frequency, 137
    low-pressure membranes, 141
    membrane bioreactor, 101–106, 203
    membranes, 35, 221
    microfiltration, 211
    physical, 106
    recovery, 94, 103–106
    recovery cleaning steps, 102, 104–106
    reverse osmosis, 169, 170, 211, 220, 221, 225
    safety, 152
    tanks, 106, 142-144
cleaning systems, 37
    automatic, 84
    chemicals, 84, 85
    chemical systems, 159
    maintenance cleaning, 84
    membrane bioreactor, 74, 84, 85
    recovery cleaning, 84
clean-in-place system, 43, 51, 52
    backpulse pumps, 83, 84
    membrane bioreactor, 74
    optimization, 136, 137
coagulation, feed water, 135
coarse-bubble aeration, 77, 205

cold weather, membranes, 193
coliforms, instrumentation, 48
colloidal material, 20
components
    membranes, 34
    low-pressure membrane systems, 123, 124, 128–130
    membrane bioreactor, 76
    posttreatment, 124, 125
    pretreatment, 126–128
    reverse osmosis and nanofiltration, 162–164
    system equipment, 40
composite asymmetric microporous membrane, 26
concentrate pressure, 244
concentrate, 16, 152, 231
concentrate, disposal, 160, 161
concentrate, disposal, staged, 162
conductivity, reverse osmosis, 173
configurations
    low-pressure membranes, 120–128
    membrane bioreactor, 80, 81
contaminants
    dimensions, 119
    conventional, 9
    emerging, 5, 9
    nonconventional, 9
contaminant removal, 17, 231, 246
    ultrafiltration and microfiltration, 117, 118
    reverse osmosis, 168
controls, 164, 167–169
controls, dose, 135
conventional activated sludge, 13
conventional contaminants, 9
conventional pollutants, 5, 9
costs, reverse osmosis, 171
crossflow
    membrane filtration, 231
    velocity, 49

*Cryptosporidium*, 20
*Cryptosporidium*, reverse osmosis, 226
cyclic aeration, 86
cylinders, spiral-wound, 25

# D

damaged membrane elements, 107
damaged membrane elements, identification and repair, 108
data analysis
    low-pressure membranes, 139, 140
    trends, 139, 140
    interpretation and reporting, 175–177
decentralized reuse, 7
dechlorination, 154–156
    low-pressure membranes, 122
    operational considerations, 131
definition
    absorption, 229
    adsorption, 229
    anion, 229
    anisotrophic membranes, 229
    array, 229
    backpulsing, 230
    backwash, 230
    binders, 230
    bubble point, 230
    cartridge filter, 230
    cassette, 230
    cation, 230
    CEB, 230
    chemically enhanced backwash, 230
    concentrate, 231
    contaminant removal, 231, 246
    crossflow, 231
    crossflow membrane filtration, 231
    diffusion, 231
    effluent, 232
    feed stream, 232
    feed water, 232
    filtrate, 232
    flat-sheet membranes, 232
    flow equalization, 232
    flux, 233, 241–243
    fouling, 233
    hollow-fiber module, 233
    immersed membrane systems, 233
    influent, 233
    integral asymmetric membranes, 234
    integrity breach, 234
    integrity monitoring system, 234
    isotropic membranes, 234
    Langelier Saturation Index, 234, 247
    macrofouling, 234
    mass-transfer coefficient, 234
    membrane bioreactor, 234
    membrane frames, 235
    membrane unit, 235
    microfiltration, 235
    microporous membranes, 235
    microfouling, 235
    module, 235
    molecular-weight cutoff, 235
    MTC, 234
    nanofiltration, 235, 236
    net driving pressure, 244
    osmosis, 236
    osmotic pressure, 236
    permeability, 236
    pore, 236
    pressure system frame, 236
    product water, 236
    programmable logic controller, 237

rack, 237
recovery, 237, 245
reject, 237
rejection, 237, 246
retentate, 237
reverse osmosis, 237–238
scaling, 238
semipermeable membrane, 238
silt density index, 238, 248, 249
skid, 238
skinned membranes, 238
solubility product, 249
solutes, 238
solvent, 238
spacers, 238
spiral-wound membranes, 239
stage, 239
Stiff and Davis Scaling Index, 247, 248
symmetric membranes, 239
thin-film composite membranes, 239
total surface area, 243
train, 239
transmembrane pressure, 239, 244
tubular systems, 239
ultrafiltration, 240
vessel, 240
denitrification, 69
denitrification, membrane bioreactor, 205
design parameters, membrane bioreactor, 196–198
differential pressure, 176, 177
diffusers
    cleaning, 106
    troubleshooting, 110
diffusion, 18, 19, 231
diffusive airflow test, 45, 46
diffusive airflow test, low-pressure membranes, 142
direct integrity testing, 45, 141
direct reuse, 6, 9
discharge, effluent, 12
disinfection, 8, 12
    low-pressure membranes, 122
    membrane bioreactor, 60, 75, 203
    operational considerations, 131
    posttreatment, 53
    reverse osmosis, 160
    UV, 75, 203
disposal, concentrate, 160, 161
dissolved material removal, microfiltration and ultrafiltration, 119
dissolved oxygen, 91, 188, 189
    membrane bioreactor, 73, 191
    meter, 88
dissolved solids, 18
dose
    control, 135
    jar test, 135
double-pass, reverse osmosis, process description, 211–212
drain pumps, 86, 87
drains, 21
drinking water standards, 8
drinking water, 2, 20, 216
drinking water, reverse osmosis, 220, 222, 227

**E**

EBPR, 187, 188, 200
effluent, 16, 232
    discharge, 12
    filtration, 28, 29, 113–146
    membrane bioreactor, 64
    membrane filtration, process flow, 121
    organic matter, 167

requirements, 5
reuse, 11, 12
reverse osmosis, 226
tertiary, 21, 149, 151
electrostatic exclusion, 18
element, 22, 25, 33, 163, 164, 232
element, low-energy, 171
EMBR, 11
emerging contaminants, 5, 9
emerging pollutants of concern, 149, 151, 159
enclosed hollow-fiber membrane, 31, 32
endocrine disruptors, 65
energy-recovery devices, reverse osmosis, 172
enhanced biological phosphorus removal, 187, 188, 200
EPOC, 149, 151, 159
equipment
    redundancy, 96, 134
    standby, 96
    troubleshooting, 109–111
equipment configurations
    components, 162–164
    low-pressure membranes, 126–128
    membrane bioreactor, 76–89
    membrane components, 128–130
    pretreatment, 126–128
    reverse osmosis and nanofiltration, 161–164
equipment manufacturers
    membrane bioreactor, 65, 66
    reverse osmosis and nanofiltration, 152
ERD, reverse osmosis, 172
esters, 27
etched polymer membrane, 26
ethers, 27
extractive membrane bioreactor, 11

## F

fats, oils, and grease removal, 39
fecal coliforms
    monitoring frequency, 139
    microfiltration and ultrafiltration, 118
feed flow, troubleshooting, 145
feed pressure, 17, 244
feed rates, tracking frequency, 138
feed stream, 16, 232
feed water, 16, 33, 153, 232
    chemical coagulation, 135
    chemicals, 124
    conductivity, 173
    low-pressure membranes, 116, 117
    microfiltration, 116, 150
    monitoring, 172–174
    monitoring frequency, 138
    quality parameters, 116
    reverse osmosis, 150
    standards, 117
    ultrafiltration, 150
    water quality, 150
fiber membrane, 21
filters
    cartridge, 39, 153, 154
    pleated-cartridge, 28
filtrate, 16, 21, 232
    flow, monitoring frequency, 138
    flow control, automated systems, 132, 133
    water quality, 117, 118
filtration, 20
    aeration, 95
    alarms, 95
    backwash, 93
    capacity management, 96
    chemically enhanced backpulse, 93
    effluent, 28, 29

equipment, 80-87
flow control, 94, 95
modes of operation, 92–95
optimization, 96, 97
recovery cleaning, 94
relax, 92, 93
secondary, 8
tertiary, 8, 9
trends, 95
fine-bubble aeration, 77, 78
fine screening, 12, 37, 66, 67, 188
    automatic, 37, 38
    low-pressure membranes, 127, 128
    operational considerations, 130–132
first pass reverse osmosis, 212
flat-plates, 25, 38, 81
flat-plates, membrane, 34
flat-sheet, 25, 27
flat-sheet, membrane, 28, 232
flow
    equalization, 37, 232
    high-flow shutdown, 48
    inside-out, 35
    meter, 133, 138
    outside-in, 35
    trends, 140
flow control, 94, 95
    automatic or manual, 48
    influent flow, 94
    setpoint, 94, 95
flux, 16, 37, 41, 42, 74, 233
    cold weather, 191
    definition, 241–243
    Key Colony Water Reuse Plant, 207
    membrane bioreactor, 189–192
    normalized, 242
    operating ranges for reverse osmosis and nanofiltration, 168
    process optimization, 135, 136
    programmable logic controllers, 48
    ranges for reverse osmosis, 164
    relax, 92, 93
    reverse osmosis, 220, 225
    setpoints, 189
    specific, 176
    system operation, 133
    temperature, 241
    transmembrane pressure, 244
    Traverse City Regional Wastewater Treatment Plant, 202
    trends, 139, 140
    troubleshooting, 110
    values for reverse osmosis and nanofiltration, 162
foam control, membrane bioreactor, 202
FOG removal, 39
foulants, reverse osmosis, 178
fouling, 17, 34, 52, 231
    barium sulfate, 166
    calcium sulfate, 166
    chemical cleaning, 177, 178
    flux, 135, 136
    macrofouling, 17
    microfouling, 17
    nonreversible, 17
    organics, 167
    pore, 17
    reversible, 17
    reverse osmosis, 157, 158, 165–167
    reverse osmosis and nanofiltration, 153
    silica, 166
    silt density index, 248, 249
    solids retention time, 72
    strontium sulfate, 166
    sulfate salts, 166

transmembrane pressure, 244
troubleshooting, 110
frame, 21–23
free chlorine, 155

## G

*Giardia lamblia*, 20
*Giardia*, reverse osmosis, 226
granular media filtration
    low-pressure membranes, 127, 128
    pore sizes, 127
grit removal, 39, 188
groundwater, 6
    recharge, 10
    reuse, 8, 214, 221
    reuse, reverse osmosis, 222
guidelines, water reuse, 8

## H

Hamptons Water Recycling Facility, 193–198
    permit requirements, 198
    plant data and design, 196–198
    process overview, 194–196
HCl, scale prevention, 156
high flow, alarms, 95
high-pressure
    inline systems, 80
    membrane systems, 18
history, timeline, 2
hollow-fiber membrane, 11, 25, 29, 32, 35, 38, 41, 81, 211, 224
    backpulse pumps, 83
    backpulsing, 93
    backwashing, 50
    diameter, 31
    microfiltration, 215
    module, 233
    outside-in flow, 30

    silt density index, 248
    submerged, 30
humic acids, 20
hydrocarbons, 27
hydrochloric acid
    recovery cleaning, 105
    scale prevention, 156
hydrogen peroxide, recovery cleaning, 105
hydrophilic, 35
hydrophobic, 27

## I

immersed membrane systems, 18, 21, 22, 25, 28, 29, 34, 35, 41, 42, 59, 80, 81, 233
    activated sludge, 58
    backwashing, 50, 51
    flat-sheet, 58
    flux ranges, 135
    hollow-fiber, 58
    low-pressure, 129
    screening, 127
    secondary clarifiers, 58
    tank cleaning frequency, 142–144
    tertiary filtration, 58
    types, 81
indirect reuse, 6, 9, 11
industrial
    applications, 214–216
    makeup water, 9
    reuse, 6, 9, 214
    pressure-driven systems, 58
influent, 16, 233
inhibitor, 211
inhibitor, threshold, 167
inline membrane systems, 80
inorganics, reverse osmosis, 218–219
in-pipe cartridge, membrane bioreactor, 58

inside-out membrane
    flow, 35
    flow, backwashing, 50
    screening, 127
    flux ranges, 135
instrumentation, 48, 164, 167–169
    calibration, 108, 109, 144, 180, 181
    feed and differential pressure, 168
    level sensors, 138
    membrane bioreactor, 87–89
    pH, 168, 169
    programmable logic controller, 88
    recovery, 167, 168
    troubleshooting, 145
    turbidimeters, 88, 89
integral asymmetric membranes, 234
integral asymmetric microporous membrane, 26
integrity
    breach, 234
    monitoring system, 234
integrity testing, 44, 106, 107
    advantages, 45–47
    bubble point test, 46
    bubble-point theory, 45
    diffusive airflow test, 45, 46
    direct method, 45, 141
    disadvantages, 45–47
    frequency, 141
    low-pressure membranes, 141–143
    particle counting and monitoring, 47
    pressure decay test, 45, 46, 107
    probing, 178, 179
    programmable logic controllers, 48
    reverse osmosis, 178–180
    sonic sensing analysis, 47
    time length of test, 142
    turbidity monitoring, 47
ions, 18, 20, 149
iron, 150, 151
    microfiltration and ultrafiltration, 118
    instrumentation, 48
    reverse osmosis and nanofiltration, 154
irrigation, 6, 8, 9
isotropic membranes, 234
isotropic pore size, 24

## J

jar testing, 220
jar testing, dose, 135
jet aeration, 77
jet aeration, air scour blowers, 86

## K

ketones, 27
Key Colony Water Reuse Plant, 4, 203–208
    flux, 207
    permit requirements, 208
    plant data and design, 207, 208
    process overview, 205–207
Keystone Colorado, 4
$K_{sp}$, 166

## L

Langelier Saturation Index, 165, 234, 247
level sensors, pretreatment, 138
life cycle cost, 32
low-pressure membranes, 18, 40, 41, 80, 81, 113–146
    backwash system, 49, 133
    backwashing frequency, 136

chemical cleaning, 52, 44, 126
clean-in-place optimization, 136, 137
components, 123, 124, 128–130
data analysis and reporting, 139, 140
effluent filtration, 113–146
equipment configurations, 126–128
feed water, 116, 117
flux, 133
flux optimization, 135, 136
fouling, 135, 136
instrumentation and controls, 48, 144
integrity testing, 45
maintenance, 140–144
membrane bioreactor, 58
operation, 128–137
polymers, 129, 130
pretreatment, 121–123
process configurations, 120–126
process optimization, 134–137
routine monitoring, 137–140
screening, 126–128
seasonal changes, 137
sludge age, 129
testing parameters, 138, 139
TMP typical variations, 133, 134
transmembrane pressure, 136
troubleshooting, 145, 146
usage, 115
LSI, 165, 234, 247

## M

MABR, 11
macrofouling, 234
maintenance cleaning, 84, 101, 102
    citric acid, 102
    sodium hypochlorite, 102
    typical steps, 102
maintenance
    chemical cleaning, 141, 177, 178
    cleaning, 101–106
    damaged membrane elements, 107, 108
    instrument calibration, 108, 109, 144, 180, 181
    integrity testing, 106, 107, 141–143
    low-pressure membranes, 140–144
    membrane bioreactor, 101–109
    reverse osmosis, 177–181, 210
    tank washing and cleaning, 142–144
manufacturer, key features, 35, 36
mass-transfer coefficient (MTC), 138, 140, 168–170, 176, 234, 241–243
membrane
    aeration, 38
    anisotrophic, 227
    applications, 9, 10–12, 26
    array, 21
    asymmetric, 24
    asymmetric microporous, 26
    backwash system, 49, 50
    cartridges, 21
    cassette, 22, 29, 30
    categories, 185
    ceramic, 26
    characteristics, 26
    chemical cleaning, 51, 52
    classifications, 18
    cleaning frequency, 221
    cleaning, 35
    clean-in-place, 51, 52
    cold weather, 193
    comparison of filtration processes, 20

compatibility to chemicals, 141
components, 15, 19, 34, 40
composite asymmetric microporous, 26
configurations, 23, 34
diameter, 34, 35
element, 22, 25, 33, 163, 164, 171
enclosed hollow fiber membrane, 31, 32
etched polymers, 26
fiber, 21
flat-plate, 34, 38
flat-sheet, 25, 27, 28
fouling, 17, 34
frame, 21–23, 235
high-pressure, 18
hollow fiber, 11, 25, 29–32, 35, 38, 41
immersed, 18, 21, 22, 25, 28, 29, 34, 35, 42, 41
inorganic, 25
inside-out, backwashing, 50
instrumentation and controls, 48
integrity testing, 45, 178–180
life cycle cost, 32
life, 221
low-pressure system, 18, 40, 41, 44, 113–146
membrane, microporous, 24
module, 21, 22, 33
nozzle, 29
operation and maintenance, 41–43
organic, 25
outside-in flow hollow fiber, 30
packing density, 32
panel, 29
pass, 22
permeability, 18, 38
permeability coefficient, 241
physical structure, 24
plate and frame, 11
pore size, 18, 24, 36, 65
pores, 26
pressure-driven, 149–181
racks, 21, 22
rotating flat plates, 25
routine monitoring, 138
semipermeable, 2
sheet, 21, 29
size-exclusion, 25
skinned, 24
spiral-wound, 25, 33
stage, 22
stretched, 26
submerged, 18
symmetric microporous, 26
temperature, 140
terminology, 16
thin film composite, 24, 163, 237
timeline, 2, 4
train, 22, 92, 93
transmembrane pressure, 36, 38
treatment history, 2
trends, 3
tubular, 11, 25, 31, 32
unit, 23, 235
membrane aeration bioreactor, 11
membrane bioreactor (MBR), 9, 11, 13, 16, 20, 28, 29, 34, 37, 40, 55–111, 234
    activated sludge quality, 72
    aeration, 77, 95
    air compressors, 87
    air scouring, 50, 51, 85, 86
    alarms, 95
    alkalinity, 63, 64, 91, 92
    applications, 59–62, 67, 68
    backpulse pumps, 83

backpulse tanks, 83, 84
backpulse, 93
backwash, 93
benefits, 59, 60
biological nutrient removal, 68, 70, 71
biological process blowers, 76, 77
biological process operation, 89–92
biological treatment, 67–73
biosolids, 75
chemical feed systems, 74
chemically enhanced backpulse, 93
cleaning systems, 74, 84, 85, 203
clean-in-place, 74
compared to CAS, 59
components, 76
costs, 62
damaged elements, 107, 108
denitrification, 69
design parameters, 192
disadvantages, 61, 62
dissolved oxygen, 73
drain pumps, 86, 87
effluent water quality, 63, 64
equipment configurations, 76–89
equipment manufacturers, 65, 66
external, 59
fat removal, 39
feed water, 150
filtration equipment, 80–87
flow control, 94, 95
flux, 189–191
foam control, 62, 202
footprint, 59
Hamptons Water Recycling Facility, 193–198

high-pressure inline, 80
influent water quality, 62, 63
in-pipe cartridge, 58
instrument calibration, 108, 109
instrumentation and controls, 87–89
integrity, 106-108
Key Colony Water Reuse Plant, 203–208
limited flow capacity, 61
low-pressure immersed, 58, 80, 81
maintenance cleaning, 101, 102
maintenance, 101–109
membrane bioreactor, relax, 42, 92, 93
membrane separation, 73, 74
mixed liquor recirculation pumps, 78–80
mixed liquor suspended solids, 72, 73, 197
mixed liquor, 63
mixers, 80
monitoring, 62, 97-101
nitrification, 68
nitrogen, 68-70, 196, 197
nonpotable reuse, 60
operation, 89–97
operational changes, 96
optimization, 96, 97
oxygen transfer, 73
permeation, 92, 93
pH, 63, 64, 91, 92
phosphorus removal, 68, 69
posttreatment, 75
pressure-driven, 58
process configurations, 66–75
recovery cleaning, 94, 103–106
recycle of mixed liquor, 70, 71
residuals disposal, 75

Running Springs Water
   Recycling Plant, 187, 189
sewer mining, 60, 61
sludge wasting, 90, 91
solids retention time, 71, 72
testing requirements, 98
timeline, 4
total suspended solids, 63
transmembrane pressure, 190
Traverse City Regional
   Wastewater Treatment Plant,
   198–203
treatment mechanism, 64
treatment range of values, 150
treatment results, 192
troubleshooting, 109–111
turbidimeters, 88, 89
turbidity monitoring, 47
typical effluent quality, 64
waste activated sludge, 75
water reuse, 203–208
membrane materials, 23, 35, 163
   advantages, 25, 27, 28
   cellulose, 27
   disadvantages, 25, 27, 28
   polyamide, 27
   polyethersulfone, 35
   polypropylene, 27, 35
   polysulfone, 27
   polytetrafluoroethylene, 27
   polyvinylidene fluoride, 28, 35
   polyvinylpyrrolindone, 35
   titanium dioxide, 28
   zirconium dioxide, 28
metals, 65
metals, reverse osmosis, 218, 219, 226
metering pumps, calibration, 131
meters
   dissolved oxygen, 88
   turbidimeters, 88
Michigan Department of
   Environmental Quality, 199

microfiltration (MF), 9, 11, 12, 13,
   18–22, 26, 28, 34, 113–146, 235
   backwash volume, 246, 247
   Chandler Arizona Reverse
      Osmosis Wastewater
      Reclamation Plant, 214–221
   chemical cleaning, 211, 224
   components, 40
   contaminants, 65
   feed water, 116, 150
   filtrate water quality, 117, 118
   instrumentation and controls,
      48
   integrity testing, 45
   key features, 35, 36
   maintenance, 140–144
   membrane bioreactor, 58
   operational considerations,
      128–137
   pretreatment, 209, 213, 221,
      223, 227
   routine monitoring, 137–140
   Scottsdale Water Campus, 224
   SDI, 210
   timeline, 4
   transmembrane pressure, 221
   treatment range of values, 150
   troubleshooting, 145, 146
   turbidity, 210
   vacuum priming system, 40
microfouling, 235
microorganisms, 9
microorganisms, pathogenic, 18
microporous membranes, 24, 235
mixed liquor
   concentration, 16
   membrane bioreactor, 63
   recirculation pumps, 78–80
   recycle, 20, 70, 71
mixed liquor suspended solids
   (MLSS), 70, 71, 90
   concentrations, 72, 73

membrane bioreactor, 194, 198, 206
monitoring, 100
troubleshooting, 110
mixers, membrane bioreactor, 80
modified Ludzack-Ettinger process, 68
module, 21, 22, 24, 33, 34, 235
molecular-weight cutoff (MWCO), 17–19, 36, 235
  low-pressure membranes, 142
  screens, 127
monitoring, 97–101, 138, 172–177
  biological treatment, 97-99
  data analysis and reporting, 139, 140, 175–177
  equipment, 100, 101
  feed water, 172–174
  frequency, 98, 175
  instrumentation, 100, 101
  low-pressure membranes, 137–140
  mixed liquor suspended solids, 100
  MLSS, 100
  parameters, 98, 138, 139
  permeability, 100
  permeate flows, 99, 100
  pressure, 175
  pretreatment, 97, 137, 138
  system performance, 99, 100
  transmembrane pressure, 100
  turbidity, 100
  systems, 37, 44
monovalent ions, 149

# N

nanofiltration (NF), 2, 11–13, 18–20, 26, 147–181, 235, 236
  applications, 149
  Chandler Arizona Reverse Osmosis Wastewater Reclamation Plant, 214–221
  chlorine, 155
  concentrate disposal, 160, 161
  configurations, 152–164
  manufacturers, 152
  flux operation ranges, 168
  flux values, 162
  fouling, 153
  postreatment, 53, 159, 160
  pretreatment, 153–158
  recovery, 245
  safety, 152
  silt density index, 248
  system components, 162–164
  timeline, 4
  transmembrane pressure, 221
  two-pass array, 163
National Pollutant Discharge Elimination System (NPDES), 5
NDMA, 12
net driving pressure (NDP), 17, 244
nitrification, 68
  membrane bioreactor, 205
  pH, 91
nitrogen
  removal, 68–70
  membrane bioreactor, 189, 192, 195–198, 203, 208
  reverse osmosis, 219
N-nitrosodimethylamine, 12
Nollet, 2
nonconventional pollutants, 5, 9
nonpotable reuse, 7
nozzle, 29

# O

one-pass reverse osmosis, 210–212
operation
  automated systems, 131–134
  costs, 171
  event changes
  fouling, 165–167

instrumentation and controls, 167–169
low-pressure membranes, 128–137
membrane bioreactor, 89–97
membranes, 41–43
parameters, membrane bioreactor, 196–198
pretreatment, 89
pressure, 19
productivity management, 133–137
reverse osmosis, 164–172
seasonal changes, 96
troubleshooting, 109–111
optimization
air scour blowers, 96, 97
chemicals, 97
pumps, 97
organic compounds, soluble, 65
organics
fouling, 167
reverse osmosis, 219
organophosphonates, reverse osmosis, 158
osmosis, 2, 3, 151, 236
osmotic
backpressure, 17
pressure, 236, 242
outside-in flow, 35
hollow fiber membrane, 30
screening, 127
oxalic acid, recovery cleaning, 105
oxygen transfer, 73

## P

packing density, 32
parameters
effluent water quality, 226
monitoring, 138, 139
reverse osmosis, 218 219

particle count, 47
data analysis, 140
monitoring frequency, 139
particulate control, reverse osmosis, 154
pass, 22
pathogenic, 9
peaking factor, membrane bioreactor, 61
percent recovery, 245
percent rejection, reverse osmosis, 176
permeability, 18, 27, 38, 236
membrane, 103
monitoring, 100
recovery cleaning, 103
temperature, 140
permeate, 16, 21, 33, 81, 82, 152, 163
degasification, 53
stabilization, 53
pressure, 17, 244
troubleshooting, 111
permeation, 92, 93
permeation, system, 41, 81, 82
permit limits, 5
pH
adjustment, 216, 217, 220
membrane bioreactor, 63, 64, 91, 92
monitoring frequency, 138, 139
reverse osmosis, 156, 168, 169, 173, 226
pharmaceutical and personal care products, 65
phosphorus
removal, 68, 69
membrane bioreactor, 189, 208
physical cleaning, 106
plate and frame membrane, 11
PLC, 40, 48, 237

PLC, membrane bioreactor, 87, 88
pleated-cartridge filters, 28
pneumatic system components, 87
point-source discharges, 3
pollutants
    conventional, 5
    emerging, 149
    nonconventional, 5
polyacrylates
    blend inhibitors, 158
    reverse osmosis, 158
polyamide, 27
polyethersulfone, 35
polymers, pretreatment, 129, 130
polymides, 163
polyphosphonates, reverse osmosis, 157, 158
polypropylene membrane, 27, 35, 211, 224
polysulfone, 27
polytetrafluoroethylene, 27
polyvinylidene fluoride, 28, 35
polyvinylpyrrolidone, 35
pore, 26, 236
    fouling, 17
    size anisotropic, 19, 24, 36, 65, 127
posttreatment, 39, 225
    alarms, 169
    components, 124, 125
    degasification, 53
    disinfection, 53, 75
    low-pressure membranes, 124, 125
    membrane bioreactor, 75
    nanofiltration, 53
    posttreatment, reverse osmosis, 53, 75
    reverse osmosis, 53, 75
    reverse osmosis and nanofiltration, 147–181
    posttreatment, reverse osmosis and nanofiltration, 147–181
    stabilization, 53
    systems, 34, 53
pressure
    differential, 168, 176, 177
    feed, 17, 168
    high-pressure shutdown, 48
    net driving, 17
    operating, 19
    permeate, 17
    reverse osmosis, 175
    transmembrane, 17
    vacuum, 41
pressure decay test (PDT), 45, 46, 107, 141, 142
pressure-driven, membrane bioreactor, 58
pressure system frame, 236
pressure vessel, 21, 25, 29, 31, 33, 34, 42,
pressurized membrane, low-pressure, 130
pretreatment, 9, 11, 40
    alarms, 169
    chemistry, 165
    chlorination/dechlorination, 122, 154–156
    components, 126–128
    level sensors, 138
    low-pressure membranes, 121–123
    membrane bioreactor, 89
    microfiltration, 209, 213, 215, 220, 222, 227
    monitoring parameters, 137, 138
    operational considerations, 129–132
    particulate control, 154
    polymers, 129, 130
    process optimization, 135

requirements, 37
requirements for primary
    clarification, 67
requirements for fine screening,
    66, 67
reverse osmosis, 172, 208, 225
reverse osmosis and
    nanofiltration, 153–158
scale inhibitors, 157, 158
scale prevention, 156–158
screening, 126–128
sludge age, 129
steps for reverse osmosis, 153
strainers, 122
system, 34
primary clarification, 67
priority pollutants, 65
process blowers, 76, 77
process configurations
    chemical cleaning systems, 159
    chemical storage and feed
        systems, 124
    concentrate disposal, 160, 161
    low-pressure membranes, 120–126
    membrane bioreactor, 66–75
    posttreatment facilities, 124,
        125, 159, 160
    pretreatment, 121–123
    residuals handling, 125, 126
    reverse osmosis and
        nanofiltration, 152–161
    tertiary treatment, 158, 159
process optimization
    backwashing frequency, 136
    chemical cleaning, 137
    clean-in-place, 136, 137
    flux, 135, 136
    low-pressure membranes, 134–137
    pretreatment, 135
    reverse osmosis, 171, 172
    seasonal changes, 137
process tanks, 21

product water, 16, 152, 236
    reverse osmosis, 160, 212
productivity management, 133–137
programmable logic controllers
    (PLCs), 40, 48, 237
    membrane bioreactor, 87, 88
protozoan cysts, microfiltration and
    ultrafiltration, 118
pumps
    automated control, 132, 133
    backpulse, 83
    calibration, 131
    drain, 86, 87
    mixed liquor recirculation,
        78–80
    optimization, 97
    permeate, 82
    sizing, 79, 80
PVDF, 35
PVP, 35

## R

racks, 21, 22, 237
recharge, 10
reclamation
    wastewater, 6
    water, 6
recovery cleaning, 84, 94, 103–106
    chemical selection, 105
    duration, 105
    factors that influence cleaning
        time, 103, 104
    typical steps, 104–106
recovery, 16, 237, 245
    permeate percentage, 245
    reverse osmosis, 167, 168, 220
    staged systems, 162
recreational reuse, 7
redundancy, 130
    equipment, 134
    reverse osmosis, 170

regulations
    concentrate disposal, 161
    reuse, 7
reject, 16, 237
rejection, 237, 246
relax, 42, 51, 92, 93, 102
residuals
    disposal, membrane bioreactor, 75
    handling, low-pressure membranes, 125, 126
    backwash water, 125
    spent cleaning solution, 125
    volume produced, 126
retentate, 16, 20, 21, 152, 237
retrofit, membrane bioreactor, 186, 187, 200, 204
reuse, 6, 12
    agricultural, 8
    decentralized, 7
    direct, 6, 9
    effluent, 11, 12
    groundwater, 8, 214, 221
    Hamptons Water Recycling Facility, 193–198
    indirect, 6, 9, 11
    indirect potable, 227
    industrial, 9, 214
    Key Colony Water Reuse Plant, 203–208
    nonpotable, 7
    recreational, 7
    regulations, 7
    Running Springs Water Recycling Plant, 186–192
    Scottsdale Water Campus, 222–227
    technologies, 7
    West Basin Water Recycling Plant, 208–214
reverse osmosis (RO), 9, 11–13, 18–20, 34, 40, 147–181, 237, 238
    aeration, 160
    alarms, 169
    alkalinity recovery, 160
    applications, 149
    biofilm, 170
    capacity management, 170, 171
    cartridge filters, 154, 210–212
    Chandler Arizona Reverse Osmosis Wastewater Reclamation Plant, 214–221
    chemical cleaning, 159, 170, 211, 225
    chloramines, 154, 155
    chlorine, 155
    cleaning agents, 177
    components, 162–164
    concentrate disposal, 160, 161
    definition, 151, 152
    double-pass, 211–212
    effluent water quality, 226
    elements, 163, 164
    equipment configurations, 161–164
    equipment manufacturers, 152
    event-driven systems, 168–170
    feed water, 150
    first pass, 212
    flux, 220, 225
    flux operating ranges and values, 162, 164, 168
    foulants, 178
    fouling, 153
    industrial applications, 214
    instrument calibration, 180, 181
    instrumentation and controls, 164, 167–169
    Key Colony Water Reuse Plant, 206
    maintenance, 177–181, 210
    monitoring, 172–177
    monovalent ions, 149

one-pass, 210–212
operation, 164–172
particulate control, 154
percent rejection, 176
pH, 156
posttreatment, 53, 159, 160
pretreatment, 153–158, 172, 208, 225
pretreatment chemistry, 165
pretreatment steps, 153
process configurations, 152–161
process optimization, 171, 172
product water quality, 151, 212
recovery, 167, 168, 220, 245
redundancy, 170
safety, 152
scale prevention, 156–158
Scottsdale Water Campus, 222–227
seasonal changes, 170
silt density index, 248
single-pass, 151, 210–212
spiral-wound, 215
stabilization, 160
staged systems, 162, 225
strainers, 210, 211
timeline, 4
transmembrane pressure, 221
troubleshooting, 180, 181
two-pass, 151, 163, 211–212
West Basin Water Recycling Plant, 208–214
rotating flat-plate membrane, 25, 34
routine monitoring
    data analysis and reporting, 139, 140
    low-pressure membranes, 137–140
    membrane system, 138
    parameters, 138, 139
    pretreatment, 137, 138
Running Springs California, 4

Running Springs Water Recycling Plant, 186–192
    design, 190, 192
    plant data, 190, 192
    process overview, 188–191

## S

safety, 120
    reverse osmosis and nanofiltration, 152
    sulfuric acid, 156
salts
    aqueous, 20
    reverse osmosis, 149, 157, 165, 211
    solubility product, 249
sampling locations, membrane bioreactor, 98
satellite systems, 6
satellite systems, membrane bioreactor, 61
SBS, 155, 156
SCADA system, 88, 100, 139 164, 173, 175, 176
scale inhibitors, 153, 157, 158
    blend inhibitors, 158
    organophosphonates, 158
    polyacrylates, 158
    polyphosphonates, 157, 158
scaling, 165, 166, 238
    barium sulfate, 166
    calcium sulfate, 166
    chemical cleaning, 177, 178
    prevention, 156–158
    reverse osmosis and nanofiltration, 153
    strontium sulfate, 166
    sulfate salts, 166
Scottsdale, Arizona, 4
Scottsdale Water Campus, 222–227

plant data, 226
process overview, 222–225
screening, 12, 39, 188, 217
    automatic, 37, 38
    bypass prevention, 39
    fine, 37
    membrane bioreactor, 191, 207
    pore sizes, 127
    pretreatment, 126–128
    size, 39
    two-dimensional openings, 39
SDI, 150, 153, 238, 248, 249
    reverse osmosis, 173, 174
SDSI, 165, 247, 248
seasonal changes, 96
    low-pressure membranes, 137
    reverse osmosis, 170
secondary clarifiers, 12
    immersed membrane, 58
secondary effluent, typical quality, 116
secondary filtration, 8
semipermeable membrane, 2, 238
separation, microorganisms, 74
setpoints
    percent recovery, 167
    chemical feed, 164
    flow, 189
    flux, 189
    recovery, 164
settleable solids, 20
sewer mining, 6, 60, 61
sheet membrane, 21
SHMP, 157, 158
sieve, 18, 19
silica, 166, 167
silt density index, 150, 153, 238, 248, 249
silt density index, reverse osmosis, 173, 174
single-pass reverse osmosis, 210–212

size-exclusion membranes, 17, 18, 25, 119
skid, 238
skinned membranes, 24, 238
sludge
    age, 129
    wasting, 90, 91
sodium bisulfite, 155, 156
sodium hydroxide, 211
sodium hypochlorite
    maintenance cleaning, 85, 102
    recovery cleaning, 105
sodium metabisulfite, 155
solids retention time, 59, 60, 71, 72, 89
    monitoring, 98
    troubleshooting, 110
solids, 18
solids, settleable, 20
solubility
    calcium carbonate, 247, 248
    Langelier Saturation Index, 165, 247
    Stiff and Davis Scaling Index, 165, 247, 248
solubility product, 166, 249
solutes, 238
solvent, 238
sonic sensing analysis, 47
source water quality analysis, 153
spacers, 29, 33, 238
spent cleaning solution, 125
spent cleaning solution, characteristics, 126
spiral-wound cylinders, 25
spiral-wound membranes, 33, 34, 239
    components, 33
    diameter, 34
    silt density index, 248
    reverse osmosis, 215
SRT, 71, 72, 89

monitoring, 98
troubleshooting, 110
stage, 22, 239
staged
    systems, 162
    reverse osmosis, 225
staging tank, 87
standards, drinking water, 8
Stiff and Davis Scaling Index, 165, 247, 248
strainers, 39
    low-pressure membranes, 127, 128
    monitoring, 135
    operational considerations, 130–132
    pretreatment, 122
    reverse osmosis, 210, 211
stretched membrane, 26
strontium sulfate, 166
submerged hollow fiber membrane, 30
submerged membrane, 18
sulfuric acid
    safety, 156
    scale prevention, 156
supervisory control and data acquisition system, 88, 139, 164, 173, 175, 176
supported liquid membrane, 26
surfactants, 211
suspended solids, 11, 18
symmetric membranes, 239
symmetric microporous membrane, 26

## T

tanks
    alarms, 95
    backpulse, 83, 84
    cleaning, 106
    drain system, 144
    process, 21
    staging, 87
    washing and cleaning, 142–144
temperature
    flux, 241
    membrane bioreactor, 190
    membrane performance, 140
    monitoring frequency, 138, 139
    reference for nanofiltration and reverse osmosis, 242
    reference for ultrafiltration and microfiltration, 242
    reverse osmosis, 173, 174
tertiary effluent, 21, 149, 151
tertiary filtration, 8, 9
    fat removal, 39
    immersed membrane, 58
tertiary treatment
    flux ranges, 164
    low-pressure membrane, 115
    process configurations, 158, 159
testing
    parameters, low-pressure membranes, 138, 139
    integrity, 178–180
thermal stability, 27
thin-film composite membranes (TFC), 24, 163, 239
threshold inhibitor, 167, 211
timeline, 2, 4
titanium dioxide, 28
TKN, 150, 151
    microfiltration and ultrafiltration, 118
    membrane bioreactor, 192, 198, 208
    reverse osmosis, 219
total coliforms, reverse osmosis, 226
total dissolved solids, 150, 151, 213
    instrumentation, 48
    reverse osmosis, 218, 226
total organic carbon, 17, 150, 151, 213

microfiltration and ultrafiltration, 118
reverse osmosis, 219, 226
total phosphorus, 150
total phosphorus, microfiltration and ultrafiltration, 118
total surface area, 243
total suspended solids (TSS), 17, 20, 63, 150, 151
    membrane bioreactor, 192, 198, 203, 208
    microfiltration and ultrafiltration, 118
    monitoring frequency, 139
trace contaminants, reverse osmosis, 75
train, 22, 239
transmembrane pressure (TMP), 17, 18, 36, 38, 176, 190, 239, 242, 244
    alarms, 48, 49, 95, 133, 134
    backwash system, 49
    backwashing frequency, 136
    changes, 140
    membrane bioreactor, 190
    microfiltration, 221
    monitoring, 100
    nanofiltration, 221
    permeate flow setpoint, 92
    recovery cleaning, 103
    relaxation, 51
    restoration, 43
    reverse osmosis, 221
    system operation, 133
    trends, 95
    troubleshooting, 145, 146
    typical variations, 133, 134
Traverse City Regional Wastewater Treatment Plant, 198–203
    permit requirements, 203
    process overview, 200–202
Traverse City, Michigan, 4
trends, 3

flow, 140
flux, 139, 140
troubleshooting
    clogs and blockages, 180
    feed flow, 145
    fouling and scaling, 180
    instrumentation, 145
    low-pressure membranes, 145, 146
    membrane bioreactor, 109–111
    reverse osmosis, 180, 181
    transmembrane pressure, 145, 146
    turbidity, 145
tubes, 25
tubular membrane systems, 11, 31, 32, 239
tubular membrane systems, diameter, 31
turbidimeters, 88, 89
turbidity, 17, 150, 151, 153
    alarms, 95
    damaged membrane elements, 107, 108
    data analysis, 140
    membrane bioreactor, 203
    membrane integrity, 106
    microfiltration and ultrafiltration, 118
    monitoring, 47, 100, 138, 139
    reverse osmosis, 173, 225–226
    troubleshooting, 111, 145
two-pass reverse osmosis, 163
    process description, 211–212

## U

U.S. Environmental Protection Agency (U.S. EPA), 3, 7, 8, 28, 91, 99
ultrafiltration (UF), 2, 9, 11–13, 18–21, 26, 28, 34 113–146, 240

backwash volume, 246, 247
components, 40
contaminants, 65
feed water, 150
filtrate water quality, 117, 118
instrumentation and controls, 48
integrity testing, 45
key features, 35, 36
maintenance, 140–144
membrane bioreactor, 58
operational considerations, 128–137
routine monitoring, 137–140
treatment range of values, 150
troubleshooting, 145, 146
vacuum priming system, 40
ultraviolet, 11
   absorption, 48
   disinfection, membrane bioreactor, 75, 203
unified atomic mass unit, 17
unit, 23
University of Cape Town Process, 200
UNR™ Process, 188, 189

## V

vacuum, 19, 25, 41
vacuum, priming system, 40
vessel, 25, 31, 33, 34, 240
vessel, pressure, 21, 29, 31
viruses, 18, 20, 150
   microfiltration and ultrafiltration, 118
   reverse osmosis, 226
viscosity of water, 242, 243

## W

waste activated sludge, membrane bioreactor, 75

wastewater reclamation, 6
Water Factory 21, 4, 10
water quality
   feed water, 150
   filtrate, 117, 118
   reverse osmosis, 151
   typical filtrate values, 118
water reclamation, 6
water reuse, 2, 6, 12
   agricultural, 8, 221
   decentralized, 7
   direct, 6, 9
   effluent, 11
   groundwater, 8, 214, 221
   guidelines, 8
   Hamptons Water Recycling Facility, 193–198
   indirect, 6, 9, 11
   industrial, 9, 214
   Key Colony Water Reuse Plant, 203–208
   nonpotable, 7
   regulations, 7
   Running Springs Water Recycling Plant, 186–192
   Scottsdale Water Campus, 222–227
   technologies, 7
   West Basin Water Recycling Plant, 208–214
water-reuse criteria, Florida, 205
water terms, 9, 16
West Basin Water Recycling Plant, 4, 208–214
West Basin Water Recycling Plant, plant data and design, 212, 213

## Z

zirconium dioxide, 28